UNIT 1

How Does the Sun Affect Earth?

UNIT 1
How Does the Sun Affect Earth?

Astronomy—space science—has both a universal and a personal face. A study of the Sun illuminates both of these faces, making solar astronomy a fitting starting point for middle school students' astronomy learning.

The universal face of astronomy is quite literal—astronomy is the study of the Universe. The visible Universe is dominated by stars, so knowledge about our local star, the Sun, is knowledge about the Solar System, other possible planetary systems, the galaxy, and the cosmos as a whole. Students see the personal face of astronomy in the effects that an astronomical object, such as the Sun, has on their daily lives. They already know that the Sun is very hot and very bright. Middle school students have most likely been taught that the Sun is the primary source of energy for Earth, that the Sun is responsible for climate and weather, and that the Sun fuels the process of photosynthesis, which feeds the Sun's energy into the living systems of Earth.

The activities in this unit maintain the personal connection to a student's life on Earth, but go way beyond the basic "very hot and very bright" concepts. Looking at the Sun as an example of a star gives students an opportunity to learn about all the energies a star can produce, including the full spectrum of electromagnetic energy and the stream of matter that flows from the Sun in the solar wind. They discover that these energies do not flow constantly from the Sun, but that dramatic bursts of energy from the Sun can race toward Earth with consequences that affect satellites, radio communications, electrical power systems, and astronaut safety. They learn about various shields that Earth has, such as the magnetosphere and the stratosphere. They also test possible shields, such as sunscreen and fabric, that reduce personal risks from solar ultraviolet energy.

Learning more about the Sun and its effects on Earth has become a critical research priority, especially in recent years. Your students will act as solar scientists to explore a mystery, create and interpret graphs, and conduct controlled experiments. By studying the fascinating and lesser-known aspects of the Sun and Earth and the connections between the two, students will see how today's fast-growing technologies have raised new questions about our vulnerability to solar storms— whether we live on Earth's surface and rely on satellite communications or venture into space beyond Earth's natural shields.

Credit: Artwork by Rose Craig based on illustrations by Steele Hill, NASA Goddard Space Flight Center.

UNIT 1

How Does the Sun Affect Earth?

SESSION SUMMARIES (8 Sessions)

1.1 The Case of the Mysterious Events

In this session, the mystery begins when events are revealed to the class through several news flashes. After recording the details of each event on a wall calendar, the class divides into teams to research the seven suspects in the case. Teams work collaboratively to move the research process forward, utilizing clues to decide which suspect to investigate next. The session concludes with a whole-class discussion about the suspects, and a chief suspect for further investigation is proposed. There are no key concepts for this session.

1.2 Scale Model of the Sun-Earth System

In this session, students begin by taking the Pre-unit 1 Questionnaire to share what they already know about the Sun. The class then briefly discusses what a star is, before moving on to construct a scale model of the Sun-Earth system. Using a 10 cm ball as a model for the Sun, the class observes how small Earth is in comparison, and just how much space separates Earth and the Sun. During this session, the key concepts that will be added to the classroom concept wall are:
- *The Sun is a star.*
- *Scientists use models to demonstrate ideas, explain observations, and make predictions.*
- *The Sun is about 150,000,000 km away from Earth.*
- *Only a tiny amount of all the matter and energy that the Sun puts out comes toward Earth.*

1.3 Energy from the Sun

In this session, students first learn about the visible spectrum before expanding their knowledge to include the full electromagnetic spectrum. The session discussion reveals that not only does the Sun radiate the full electromagnetic spectrum, but it also puts out the solar wind, a constant stream of energetic particles. The understanding that students gain in this session of what is emitted on a "normal" day for the Sun prepares them for future sessions when they will learn about solar storms on the Sun and their effects on Earth. During this session, the key concepts that will be added to the classroom concept wall are:
- *The visible light our eyes can detect is part of a larger spectrum of electromagnetic energy.*
- *The Sun radiates the full spectrum of electromagnetic energy.*
- *The solar wind is a constant stream of charged particles that the Sun puts out.*

1.4 A Stormy Sun: Revisiting the Mystery

In this session, students learn that the Sun has periods of intense activity when solar flares and coronal mass ejections can occur. Drawing from their knowledge of solar science, student teams discuss and write answers to questions about the mystery events in preparation for their roles as "expert witnesses" in a trial implicating the Sun as the culprit in the mystery. The students' "testimony" brings out the intriguing match between the dates of the disruptions on Earth and the earlier occurrence of a huge solar storm. During this session, the key concepts that will be added to the classroom concept wall are:
- *Solar flares and coronal mass ejections (CMEs) occur during solar storms, when the Sun is active.*
- *A solar flare releases large amounts of electromagnetic energies and solar particles into space.*
- *A CME ejects particles and material from the Sun's corona at high speeds into space.*
- *Particles released by solar-storm events—such as solar flares and CMEs—travel much more quickly than particles in the solar wind.*
- *The amount of particles and energy put out by the Sun is not constant.*
- *Scientific explanations are based on evidence gathered from observations and investigations.*

1.5 The Balloon-Rocket Mission

In this session, student teams work together to graph and compare data obtained from the balloon-rocket mission. Analysis of the completed graphs shows that shields at various altitudes prevent some solar energies from reaching Earth's surface. The session ends with a class discussion about Earth's magnetosphere and its role as a shield against solar particles. During this session, the key concepts that will be added to the classroom concept wall are:
- *Shields at various altitudes prevent some solar energies from reaching Earth's surface.*
- *The magnetosphere shields Earth from solar particles at a very high altitude.*
- *The changing shape of Earth's magnetosphere during a solar storm can leave satellites and astronauts unshielded.*

1.6 Investigating Ultraviolet Shields

In this session, the class learns more about ultraviolet energy and its hazards. Using UV beads, which change color when exposed to ultraviolet light, student teams design and carry out experiments to test various shield materials. The importance of using a control and taking careful observation notes is emphasized as students conduct their tests. During this session, the key concepts that will be added to the classroom concept wall are:
- *Everyone, especially those at high altitudes, should be concerned about ultraviolet radiation from the Sun.*
- *Earth's ozone layer shields us from some of the Sun's ultraviolet energy.*

1.7 Concluding the Ultraviolet-Shields Investigation

In this session, teams begin by discussing their experiment successes and mistakes with one another. Students are encouraged to learn from one another's mistakes and are given additional time to continue their shielding experiments. The session ends with a class discussion of experiment results, and students are asked to assess the ultraviolet-exposure risk of four people located at different altitudes. During this session, the key concept that will be added to the classroom concept wall is:
- *Various materials can shield a person from UV radiation, but some shields are more effective than others.*

1.8 Living with a Stormy Sun

In this session, students review what they have learned about the possible terrestrial effects of a solar storm. A student reading about astronauts on the International Space Station teaches students about how astronauts protect themselves from exposure to dangerous energies from the Sun. With the case of the mysterious events finally solved, students finish the unit by taking the Post-unit 1 Questionnaire. During this session, the key concept that will be added to the classroom concept wall is:
- *We must all be concerned about bursts of energy from the Sun.*

SESSION 1.1

The Case of the Mysterious Events

Overview

This first session opens with a mystery that sets the stage for this unit on solar science. Mysterious events occurring over a span of a few days are revealed to the class. Some of these events, such as disrupted communications and power blackouts, have serious consequences for individuals and communities in different parts of North America. In subsequent sessions, as students learn more about the Sun (and some of its lesser-known, yet important, effects), students are invited to apply their newfound knowledge to explain the mysterious events of this first session. In this session, the mystery begins when events are revealed to the class through several news flashes. After recording the details of each event on a wall calendar, the class divides into teams to research the seven suspects in the case. Teams work collaboratively to move the research process forward, utilizing clues to decide which suspect to investigate next. The session concludes with a whole-class discussion about the suspects, and a chief suspect for further investigation is proposed. There are no key concepts for this session.

The Case of the Mysterious Events	Estimated Time
Unsettling News Flashes	15 minutes
Researching Possible Suspects	20 minutes
Deciding on a Chief Suspect	10 minutes
Total	**45 minutes**

What You Need

For the class:
- ❏ overhead projector or computer with large-screen monitor or LCD projector
- ❏ (optional) Space Science Sequence CD-ROM
- ❏ 1 copy of News Flashes for the Mysterious Events student sheet (two pages) from the copymaster packet or CD-ROM file
- ❏ a large sheet of butcher paper
- ❏ a wide felt-tipped marking pen
- ❏ transparency of the Satellite Information Sheet from the transparency packet or CD-ROM file

For each team of 2–3 students:
- ❏ 1 copy of each of the following seven Information Sheets (one page each): Satellite, Weather, Unusual Lights in the Sky, Nuclear Weapons, Solar Wind, Solar-Surface Activity, Earth's Magnetosphere from the copymaster packet or CD-ROM file

Unit Goals

The Sun is a star, and a main source of energy for Earth.

The Sun gives off the full spectrum of electromagnetic energies, as well as solar particles.

The Sun's energy and matter output varies and is not constant.

Earth has protective shields located at various altitudes that help to block much of the Sun's harmful output from reaching Earth's surface.

Safety is a concern — without Earth's shields to protect us, some of the Sun's energies can be harmful.

TEACHER CONSIDERATIONS

TEACHING NOTES

Students love mysteries! Use every opportunity to play up the mystery angle by using words such as *clue, lead,* and *suspect* to generate and maintain your students' interest. The mystery in this session drives the rest of the unit, so make an effort to "hook" your students during this activity!

Pre-unit questionnaires are taken in the first session for the other units in this sequence. For this unit, however, the mystery is introduced first, and students take the Pre-unit 1 Questionnaire in Session 1.2.

Key Vocabulary

Scientific Inquiry Vocabulary

Control
Evidence
Model
Observation
Prediction
Scale
Scale model
Scientific explanation

Space Science Vocabulary

Coronal mass ejection (CME)
Electromagnetic (EM) energy
Magnetosphere
Matter
Particle
Shield
Solar flare
Solar particle
Solar wind
Spectrum
Star
Ultraviolet (UV)

SESSION 1.1   The Case of the Mysterious Events

For each student:
- ❏ scratch paper
- ❏ pencil
- ❏ 1 copy of the Research Notes student sheet (two pages) from the copymaster packet or CD-ROM file

Getting Ready

1. **Arrange for the appropriate projector format to display images to the class.** Decide whether you will be using the overheads or the CD-ROM. Set up an overhead projector or a computer with a large-screen monitor or LCD projector.

2. **Prepare the news-flash strips.** Make a copy of the News Flashes for the Mysterious Events student sheet and cut the news flashes into 14 separate strips.

3. **Prepare the week-long calendar.** On a large sheet of butcher paper, draw out a week-long calendar, with April 20 through 26 labeled, as shown below. Post it somewhere in the classroom where you can easily write on it, and where all students will be able to see it.

April 20	April 21	April 22	April 23	April 24	April 25	April 26

4. **Decide how you will divide the class into teams of 2–3 students.**

5. **Copy and stack the Information Sheets.** Copy all seven Information Sheets, making one copy of each sheet for each team of 2–3 students. Stack the Information Sheets according to topic (e.g., Satellite, Weather, etc.) and place the stacks on a table or counter in the classroom where they will be easily accessible by the students.

6. **Make copies of the Research Notes student sheets.** Make a copy for each student.

7. **Familiarize yourself with the mystery.** Go through the news flashes and read the solution to the mystery on page 153.

8. **Optional:** If you plan to use the Space Science Sequence CD-ROM, set up the computer with large-screen monitor or LCD projector. The CD-ROM can be used to show news-flash graphics to the class.

TEACHER CONSIDERATIONS

TEACHING NOTES

If you are short on time, fill in the calendar with pre-written key points, cover these up, then "reveal" them as students bring them up during the discussion.

Several events may fall on the same dates, so be sure to leave enough room on the calendar for each news flash. Dates with multiple events: April 21 (3 events); April 23 (5 events); April 24 (3 events); April 25 (2 events). (NOTE: Be sure to also save the calendar, as it will be used again throughout the unit.)

SESSION 1.1 The Case of the Mysterious Events

News Flashes for the Mysterious Events

1. **News Flash: April 21**
It was a tense and fearful hour for a California snowboarding club earlier today. Some of their club members were on an extreme snowboarding expedition on Mount Everest, halfway around the world. The California club members were communicating with the expedition by shortwave radio when suddenly, at 3:18 p.m. Pacific Daylight Time, all radio signals were interrupted. At first, club members feared that the expedition may have encountered an avalanche or some other mishap, but they were relieved to learn that everyone in the expedition was safe when radios began to function normally about an hour later.

2. **News Flash: April 21**
Air traffic controllers in Chicago reported that fifteen airplanes in various parts of the country lost radio contact for about an hour. The problem occurred at 3:18 p.m. Pacific Daylight Time and continued until 4:20 p.m. Pacific Daylight Time.

3. **News Flash: April 21**
International Space Station radiation sensors showed very high levels of X-rays and gamma rays earlier today. The unusual radiation was detected at 3:18 p.m. Pacific Daylight Time. Sources report that analysts in Washington D.C.'s Defense Department spent the rest of the evening investigating the incident.

4. **News Flash: April 23**
ABZ television broadcasts were disrupted today for two hours.

5. **News Flash: April 23**
A group of college science students and professors working in California's Mojave Desert witnessed very unusual clouds of light and colors in the sky in the direction of Los Angeles. They immediately suspected a nuclear explosion in Los Angeles.

6. **News Flash: April 23**
A huge power surge caused electrical blackouts in New York, Boston, and Seattle. Power was restored to Boston after four hours and to Seattle after six hours.

GO! Unsettling News Flashes

1. **Mysterious events are happening.** Tell the class that they are about to be presented with a series of strange and mysterious events. Explain that although the scenario itself is fictional, the events described are things that have happened, or could happen, in real life.

2. **Students will be scientific detectives.** Explain that the class will be listening to a series of news flashes read by student "newscasters." Students should pay careful attention to each news flash and try to pick up clues about each event. Tell them that later they will work in teams as "scientific detectives" to research potential "suspects" as they try to determine what is causing the mysterious events!

3. **Recording events on the wall calendar.** Show students the wall calendar and tell them that it will be used to keep track of important information for each event. After each news flash has been read, students should list any clues they've gathered for you to record on the wall calendar.

4. **Encourage students to ask questions.** Pass out a sheet of scratch paper and a pencil to each student. Tell students that they will most likely have many questions as the news flashes are being read. Encourage students to record their questions on the scratch paper.

5. **Select a student to read the first news flash.** Ask for a student volunteer to be the first "newscaster." Hand the student the first news-flash strip to read.

6. **Ask students what the important points were in the first news flash.** Have students summarize any important points or clues for you to record on the calendar. If no one suggests that the time of the event is significant, mention that it may be noteworthy and worth recording.

7. **Continue reading and recording information for each news flash.** Select additional students to read the remaining news flashes *in sequence*. After each news flash is read, have the class summarize the important points of each event for you to record on the class calendar.

TEACHER CONSIDERATIONS

TEACHING NOTES

For more advanced students, you can make the mystery more difficult by giving the times of the events in different time zones.

3:10 Pacific Daylight Time (or 15:10 PDT) is equivalent to:
- 4:10 Mountain Daylight Time (or 16:10 MDT)
- 5:10 Central Daylight Time (or 17:10 CDT)
- 6:10 Eastern Daylight Time (or 18:10 EDT)
- 22:10 Universal Time

What is Universal Time? Astronomers describe events in Universal Time—an international standard time that is the same for all time zones around the world. It is based on local standard time at the Greenwich Observatory in England.

Discussing time zones. If you provide times of the events in the mystery in different time zones, the activity offers a good opportunity for students to share what they know about local time zones. After students examine the news flashes, they should come to the conclusion that some events occurred at the same time.

SESSION 1.1 The Case of the Mysterious Events

8. **More information to address unanswered questions later.** As each mysterious event is revealed, encourage students to ask questions. If there is enough information in the news flash to answer a student's question, the class should do so. Otherwise, remind students to record their questions on their sheets of scratch paper. They will be able to follow up on them later when they have more information.

9. **Ask students if there are any patterns in the sequence of events.** After all the news flashes have been read, ask the class if they've noticed any pattern in the sequence of events. You might ask which times or days had many things happening and which times or days had no mysterious events occurring.

10. **The class discusses what might be happening.** Call on students to share their ideas about how these events might be related to one another or what might be causing them. Let the class know that they will be researching possible causes of these mysterious events next.

TEACHER CONSIDERATIONS

TEACHING NOTES
Rather than telling students their questions will be "answered" later, the strategy of encouraging students to record their questions to keep in mind as they gain more information is an excellent way to strengthen the inquiry aspects of the activities. It stimulates further thinking and research and empowers students as independent learners and inquirers.

FOR TEACHERS ONLY: THE SOLUTION TO THE MYSTERY
Important: Do not reveal to students, until the end of this activity, that this is a unit about the Sun!

Later in this unit, students will learn the Sun-Earth connection behind the mystery. Here's a quick overview for you now, so that you can better guide your student detectives during this first activity.

At 3:10 p.m. Pacific Daylight Time (PDT) on April 21, a huge solar flare occurred. A *solar flare* is a temporary brightening of a portion of the Sun's surface with a significant increase in X-ray and gamma-ray energy. Eight minutes later (the time it takes light to travel from the Sun to Earth), the flare was observed from Earth. An increase in X-rays and gamma rays from the flare also arrived at 3:18 p.m. PDT and caused some of the strange events that day: military satellites were alerted to increased levels of X-rays and gamma rays; radio transmissions were interrupted as the X-rays and gamma rays disturbed the upper atmosphere of Earth. The solar flare was accompanied by a huge gust of particles from a coronal mass ejection (CME), which erupted from the Sun's atmosphere, blowing out some of the gas from the corona in the direction of Earth. CME particles travel more slowly than the speed of light—it took the main gust of particles two days to reach Earth. When the CME swept past Earth on April 23, it interacted with Earth's magnetosphere and caused many of the events that took place that day, including the crippling of several satellites, a power surge and blackouts, and dramatic auroras in the night sky.

CD MULTIMEDIA NOTES
News Flashes for the Mysterious Events. Use these news-flash graphics on the CD-ROM to supplement the student-read "news flashes." Each news flash appears chronologically on a separate screen. Advance or rewind the interactive by clicking the NEXT or BACK buttons, respectively, in the lower right-hand corner of the screen. To enlarge the interactive to full screen, click CONTROL F for Windows and APPLE F for Macs. Click ESC to exit this mode. (You can close the program any time, just like you would close a window on your desktop.)

SESSION 1.1 The Case of the Mysterious Events

Researching Possible Suspects

1. Students will work in teams to solve the mystery. Divide the class into teams of 2–3 students. Say that each team will work together to try to determine what might be causing the mysterious events. Explain that in order to solve the mystery, teams will first have to learn more about some of the possible causes, or suspects, in the case.

2. Go over the case's suspects with the class. Pass out the Research Notes student sheets and briefly go over the seven listed suspects with the class. Point out if any of the suspects have already been mentioned by students in the previous discussion.

3. Explain how teams should use the Information Sheets to conduct their research. Tell the teams that each suspect has an Information Sheet, which they should use to look for useful information, or clues, about the suspect. Remind teams that they are trying to solve a mystery, so useful information would be anything that might help them to figure out which suspect(s) caused the mysterious events reported in the news flashes.

4. Teams may look at only one Information Sheet at a time. Point out where you've placed the Information Sheets for each suspect. Explain that each team should decide together which suspect they would like to investigate first. One team member should then retrieve the suspect's Information Sheet and bring it back for the team to look at together.

5. Each student should make his or her own research notes. Students should use their Research Notes student sheets to keep track of any clues they find for each of the suspects. Let them know they should not write on the Information Sheets. Emphasize that although the research will be done in teams, each student will be responsible for taking his or her own notes. After everyone on the team has made their notes, the team decides together which suspect to research next. One team member should return the first Information Sheet and retrieve the next one for the team.

TEACHER CONSIDERATIONS

TEACHING NOTES
Research notes are critical tools for scientists. Their notes are a record of their work—their methods and procedures, data, conclusions, and new questions for further research. Students will refer to their notes at the end of this session and again in Session 1.4 as they confirm that the Sun is the culprit in the mystery.

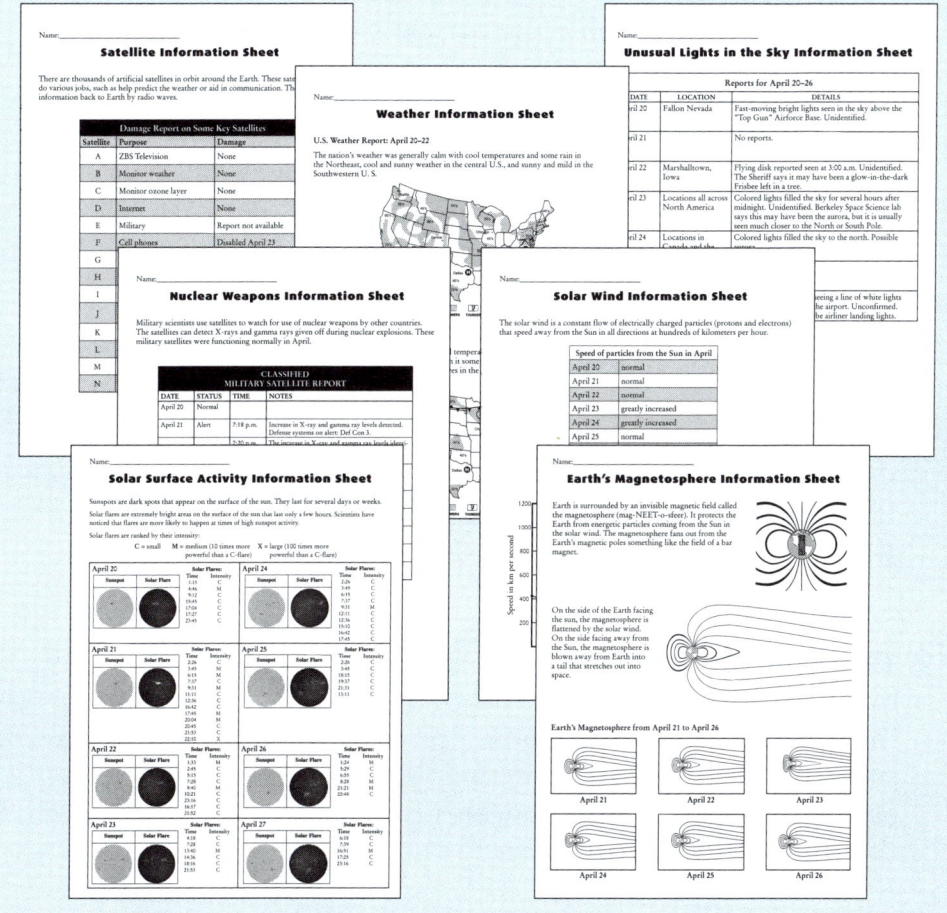

SESSION 1.1 The Case of the Mysterious Events

6. **Model the research process with the class using the Satellite Information Sheet.** Go through the following steps with the class:

   a. **Show the transparency of the Satellite Information Sheet.** Give the class a minute to look over the sheet's contents.

   b. **Ask, "What kind of information does the sheet provide?"** [It's a report with information about damage to key satellites.]

   c. **Ask, "Does this sheet provide any information that might be a useful clue or lead to solving the mystery?"** Students may point out several things. [Satellite F, a cell-phone satellite, was disabled April 23, etc.] For each answer given, ask how it might be connected to the mysterious events.

   d. **Ask, "Do you need to write all the information from this sheet in your research notes?"** [No, only helpful information for solving the mystery should be written down.] Students should realize that not every piece of information given will be useful in unraveling the mystery. Ask the class for some examples of useful things from the Satellite Damage Report they can put on their Research Notes sheets. [Satellite K, the ABZ Television satellite, was disabled on April 23, etc.]

   e. **Teams should use what they've learned to decide which suspect to research next.** Tell teams they should consider clues gained from one suspect's information sheet to help them decide which suspect to research next. For example, after looking through the Satellite Damage Report, students may want to research a suspect they think could cause damage to satellites.

7. **Research should be conducted collaboratively as a team.** Emphasize that when they are finished with an Information Sheet, the team decides *together* which suspect to research next. Team members should aim to work collaboratively, much like scientists often do.

8. **Everyone on the team should participate.** Say that teams should work together in a way that allows everyone to participate. For example, students could take turns reading the Information Sheets aloud for the team to discuss. Remind students that although they are working in teams, everyone is responsible for writing down his or her own evidence and ideas on their Research Notes student sheets.

TEACHER CONSIDERATIONS

TEACHING NOTES
Use this opportunity to clarify the purpose of the Information Sheets with the class.

SESSION 1.1 The Case of the Mysterious Events

9. **Teams should research as many causes, or suspects, as possible.** Tell teams that they should research as many suspects as they can in the time available. Emphasize that just as important as identifying potentially guilty suspects is ruling out any suspects *not* responsible for the mysterious events!

10. **Teams might find it helpful to refer to the posted calendar of events.** Remind teams that the posted calendar with notes about each of the mysterious events may be useful to refer to as they conduct their research.

11. **Research brings up questions.** Make sure that teams understand they don't need to use the Information Sheets in any particular order—just an order that makes sense to the team. Explain that the process of research often brings up new questions, and the team should use their questions to help them decide what to research next.

12. **Teams begin their research of the suspects.** Have teams start their research and circulate among them, offering advice as necessary. Give students a five-minute warning before the end of the research session.

Deciding on a Chief Suspect

1. **Scientists often work collaboratively.** Regain the attention of the entire class. Tell teams that the collaborative research they have been doing is similar to how scientists will often approach a problem in their field of study. They work in teams and share information with one other. They also check each other's work and sometimes disagree.

2. **Survey the class.** Ask for a show of hands from teams who think they know what might have caused the mysterious events to occur.

3. **Discuss, as a class, one of the suspects.** Start with one of the suspects unrelated to the Sun, such as the weather or nuclear weapons. Invite members from one team to share their thoughts about how the suspect might have caused or be related to the mysterious events, and then have the class discuss the team's reasoning. Allow students who disagree to voice their opinions as to why they disagree. Decide as a class whether or not the suspect can be ruled out as the cause of the events.

TEACHER CONSIDERATIONS

TEACHING NOTES

If time is an issue, make sure all suspects get researched by assigning teams to different ones. Another option would be to require all teams to research all seven topics. (Be aware that this option is much more time intensive!)

One teacher said, "Over half of the class needed further time to research. I was surprised to find out that they did not figure out that the unit was about the Sun. They loved the mystery."

SESSION 1.1   The Case of the Mysterious Events

4. **Discuss the remaining suspects.** Continue through all the remaining suspects. It may be necessary to conduct some impromptu class votes to reach agreement on whether or not to rule out a possible suspect. Throughout the discussion, ask questions to force students to critically evaluate whether or not a piece of evidence they have (for or against a suspect) is strong enough to support their arguments.

5. **Suggest the Sun as a likely suspect.** After all suspects have been discussed, bring up the possibility that the Sun may be worth investigating further. While not all students may conclude that the Sun is the chief culprit, most will agree that the Sun is worthy of further study. Ask the class if they find it surprising that the Sun could cause such disastrous effects.

6. **Students will learn more about the Sun in this unit.** Tell students that while they may not have been familiar with all the suspects (such as the solar wind or solar-surface activity), they are about to begin a unit on solar science, during which they will learn more about these topics. Students will learn why it is important to understand how the Sun affects us here on Earth.

7. **The Research Notes student sheets will be used during Session 1.4.** Students will need to refer to their research notes in Session 1.4. Decide whether you would like to collect the student sheets or have students hold onto them for use later.

8. **Save the week-long calendar of mysterious events for use again in Session 1.4.** Students will need to refer to the calendar as they prepare to become "expert witnesses" in solar science.

TEACHER CONSIDERATIONS

TEACHING NOTES

About evidence. In Session 1.5, students will gather their own evidence from a mock balloon-rocket mission and define *scientific evidence* as something that can be observed, discussed, and verified by scientists and then used in forming an explanation. For now, you could briefly define *evidence* as information or clues related to the mystery. You might want to ask:

- What makes evidence strong?
- Is the evidence well researched?
- Does it come from a reliable source?
- Is there only one occurrence of the evidence, or are there multiple examples?

These kinds of questions encourage critical thinking in your students and can help them to understand the process of evaluating evidence.

Reassure students that they will learn more about solar flares and coronal mass ejections (CMEs) in Session 1.4.

SESSION 1.2

Scale Model of the Sun-Earth System

Overview

In this session, students begin by taking the Pre-unit 1 Questionnaire to share what they already know about the Sun. The questions are designed to tease out student understanding of the Sun and its effects on Earth. Question #6 asks students to draw the Sun-Earth system, representing the sizes and distances of both bodies as accurately as they can. The class then briefly discusses what a star is before moving on to construct a scale model of the Sun-Earth system. Using a 10-cm ball as a model for the Sun, the class observes how small the Earth is in comparison, and just how much space separates Earth and the Sun. During this session, the key concepts that will be added to the classroom concept wall are:

- *The Sun is a star.*
- *Scientists use models to demonstrate ideas, explain observations, and make predictions.*
- *The Sun is about 150,000,000 km away from Earth.*
- *Only a tiny amount of all the matter and energy that the Sun puts out comes toward Earth.*

Scale Model of the Sun-Earth System	Estimated Time
Taking the Pre-unit 1 Questionnaire	20 minutes
Focusing on the Sun	5 minutes
A Scale Model of the Sun-Earth System	20 minutes
Total	**45 minutes**

What You Need

For the class:
- ❏ overhead projector or computer with large-screen monitor or LCD projector
- ❏ prepared key concept sheets from the copymaster packet or CD-ROM file
- ❏ two sentence strips
- ❏ a marker
- ❏ a ball about 10 cm in diameter (a balloon or a softball will work as well)
- ❏ transparencies of the Pre-unit 1 Questionnaire (three pages) from the transparency packet or CD-ROM file
- ❏ (optional) a meter stick

For each student:
- ❏ 1 copy of the Pre-unit 1 Questionnaire (three pages) from the copymaster packet or the CD-ROM file

Unit Goals

The Sun is a star, and a main source of energy for Earth.

The Sun gives off the full spectrum of electromagnetic energies, as well as solar particles.

The Sun's energy and matter output varies and is not constant.

Earth has protective shields located at various altitudes that help to block much of the Sun's harmful output from reaching Earth's surface.

Safety is a concern — without Earth's shields to protect us, some of the Sun's energies can be harmful.

TEACHER CONSIDERATIONS

TEACHING NOTES

This session may require more time. Prepare for an extra session or move on to the scale-model activity if your students finish their questionnaires before the suggested 20 minutes.

Use the transparency of the Pre-unit 1 Questionnaire to review Question #6 with your students *after* completing this session. Be sure to save the transparency for use again when making additional questionnaire connections later on in the unit.

Key Vocabulary

Scientific Inquiry Vocabulary

Control
Evidence
Model
Observation
Prediction
Scale
Scale model
Scientific explanation

Space Science Vocabulary

Coronal mass ejection (CME)
Electromagnetic (EM) energy
Magnetosphere
Matter
Particle
Shield
Solar flare
Solar particle
Solar wind
Spectrum
Star
Ultraviolet (UV)

SESSION 1.2 Scale Model of the Sun-Earth System

Getting Ready

1. **Designate a wall or bulletin board in the classroom as the unit's concept wall.** Use the concept wall as a space to post the key concepts for each session in this unit. It is helpful to distinguish between the two categories of key concepts—Space Science and Scientific Inquiry. See page 163 for an example of a concept wall. Although the concept wall does take up a fair amount of space, it is a very useful and important learning tool for your students.

2. **Prepare the key concept sheets.** Make a copy of each key concept and have them ready to post onto the classroom concept wall during the session.

3. **Prepare the concept wall headings.** Using the sentence strips and marker, make the two headings for the concept wall—one labeled Key Space Science Concepts and the other labeled Key Scientific Inquiry Concepts.

4. **Make a copy of the Pre-unit 1 Questionnaire for each student.**

5. **Decide where you will conduct the scale-model activity.** Read through the scale-model activity as detailed on pages 168–172 and decide whether it will be possible to construct the scale model in the classroom. If not, find a space outdoors (or in the hallway or a multi-purpose room) for the activity.

GO! Taking the Pre-unit 1 Questionnaire

1. **Remind students of the mysterious events from Session 1.1.** Ask the class to think back to the strange news flashes from the previous session. Remind students that at the end of the session, the Sun was suggested as a likely suspect for the cause of the mysterious events.

2. **Introduce the questionnaire.** Tell the class that they will be learning more about the Sun in this unit. But first, they will take a questionnaire that will give them the opportunity to share some ideas they have about the Sun. Emphasize that the questionnaire is not a test, but a chance for them to start thinking about the Sun and what they might want to learn about it.

3. **Stress individual responses and independent work.** Tell students they should work silently and without helping others. They will have an opportunity later to discuss the questions and share their ideas with one another.

TEACHER CONSIDERATIONS

TEACHING NOTES

The key concepts can be posted in many different ways. If you don't want to use sentence sheets, here are some alternatives:

- Write the key concepts out on sentence strips.
- Write the key concepts out before class on a posted piece of butcher paper. Cover each concept with a strip of butcher paper and reveal each one as it is brought up in the class discussion.

A large classroom might be close enough to 12 m (40 ft) if measured from corner to corner, diagonally. Note: If it's not practical for students to leave the classroom, prepare by pacing off about 40 feet from a point in your classroom to another point down the hall or in the schoolyard that is visible through the classroom windows. If possible, choose a direction in which some known landmark (such as the flagpole or a drinking fountain) is about 40 feet away. Make note of a few other landmarks in the school that are approximately 40 feet away.

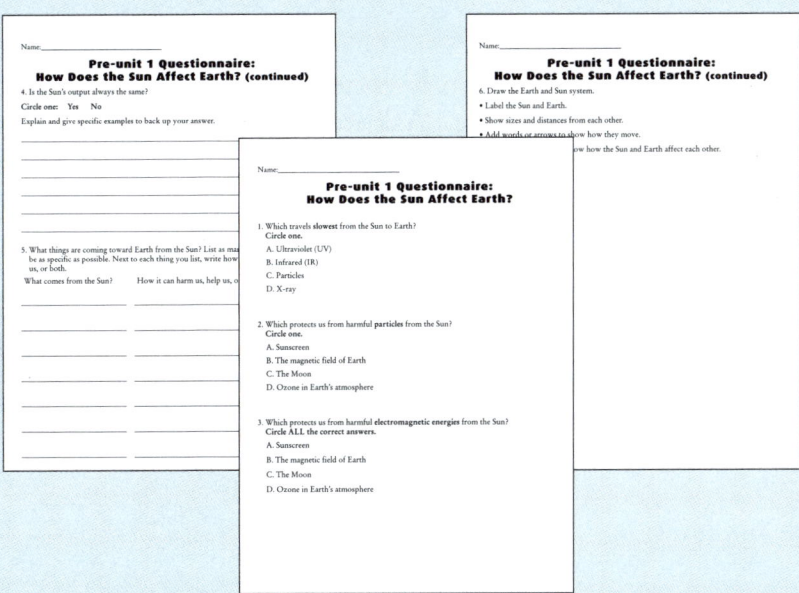

SESSION 1.2 Scale Model of the Sun-Earth System

4. **Go over Question #6.** Tell students that the last question on the questionnaire will ask them to draw a diagram of the Sun-Earth system. Say that they should follow the instructions and include as much information as they can in their diagram.

5. **Pass out the Pre-unit 1 Questionnaire.** Tell the class that they will have 15 minutes to fill out the questionnaire. When 5 minutes remain, ask students to begin work on their Sun–Earth diagrams, if they haven't already done so.

6. **Collect the questionnaires.** After every student has had a chance to answer Question #6, collect the questionnaires.

7. **Tell the class that they'll take the questionnaire again.** Explain to students that after several more class sessions studying the Sun, they will take the same questionnaire again. As they learn more about the Sun in this unit, they should think about how they might answer the questions on the questionnaire differently or more completely later on.

Focusing on the Sun

1. **What is a star?** Ask if students have heard that the Sun is a star. (Most students probably have.) Ask students what a star is. [A very large, hot, bright ball of gases.] To give students a sense of the immense size of the Sun, tell them that more than 99% of the mass of the entire Solar System is found in the Sun.

2. **Introduce the concept wall.** Point out the space in the classroom you've designated for the concept wall. Explain that as the class learns important or key concepts about the Sun, the concept wall will help them to keep track of what they've learned. Post on the concept wall, under Key Space Science Concepts:

   The Sun is a star.

3. **Discuss why the Sun is important to us.** Ask students why the Sun is important to us on Earth. [It provides the light energy that warms Earth and supports all life. Without the Sun, no life would exist on Earth!]

4. **Why the Sun looks so much larger and brighter than other stars.** Ask students why they think the Sun looks larger and brighter than other stars. Explain that since the Sun is much closer to us than any other star, it appears larger and brighter than any other star.

TEACHER CONSIDERATIONS

ASSESSMENT OPPORTUNITY
CRITICAL JUNCTURE: NECESSARY PREREQUISITE CONCEPTS
In order for students to understand the scale model of the Sun-Earth system in this session, they should know that Earth and Sun are both spherical, and that Earth orbits the Sun. If student answers to Question #6 on the Pre-unit 1 Questionnaire reveal that they do not understand these necessary prerequisite concepts, we strongly recommend that you review these with them before proceeding with this unit. These concepts are taught for full understanding in the *GEMS Space Science Sequence for Grades 3–5*.

One teacher said, "I was amazed at how little my classes understood about the Sun-Earth system and it was eye opening to me to see their questionnaire responses. The most informative question was #6 where the students needed to draw the system."

SESSION 1.2 Scale Model of the Sun-Earth System

A Scale Model of the Sun-Earth System
Scale Model: Comparative Sizes of the Sun and Earth

1. **Show the class the model of the Sun.** Tell the class that one of the first things they are going to learn about the Sun is how big and how far away it is—relative to Earth's size and location. Show students the 10-cm ball and tell them that since a star is much too big to bring into the classroom, the class is going to use the ball as a model of the Sun instead.

2. **Discuss the idea of a model with students.** Explain that a model is not the real thing, but shows something about the real thing—such as what it looks like or how it moves. Tell students that even the diagrams they drew for the last question on the questionnaire are models (but in two dimensions instead of three). Explain that scientists frequently use models to observe, study, and understand something. Post on the concept wall, under Key Scientific Inquiry Concepts:

    Scientists use models to demonstrate ideas, explain observations, and make predictions.

3. **Explain the purpose of this model.** Tell students that the purpose of the model they are about to study is to help them understand the sizes and distances in the system of the Sun and Earth. The model will be a scaled model, which means that if the real Earth is smaller than the real Sun, this model Earth will be smaller, in the same proportion, than the model Sun.

4. **Discussing how big the model Earth should be.** Hold up the 10-cm ball and remind the class that in this model the ball represents the Sun. Discuss with the class how big the model of Earth should be compared to this 10-cm model of the Sun. (If time permits, take a few guesses from the students.)

5. **In this model, a pencil point represents Earth.** Find a student with a sharp pencil and have him or her hold the pencil point up for the class to see. The point of a sharp pencil is about 1 mm across—approximately one one-hundredth of 10 cm, the diameter (or distance across) of the model Sun. Explain that the diameter of the real Sun is about 100 times the diameter of the real Earth. (It's actually closer to 109 times.) Say that compared to the model Sun, the Earth would be a sphere about the size of the pencil point. Make sure students understand that the Earth's shape is actually spherical.

TEACHER CONSIDERATIONS

TEACHING NOTES
INTRODUCING THE CONCEPT OF MODELS

If your students have not been introduced to models, consider spending some extra time explaining models and their use by scientists as in the excerpt below from Unit 1, Session 4 of the Grades 3–5 portion of the GEMS Space Science Core Curriculum Sequence.

Introduce models. Hold up a matchbox car (or any other model) and say that a model is something that shows or explains what the real thing is like. Scientists use models to show how something works and to learn more about things that are hard to study firsthand.

All models are different from the real thing in at least one way. Emphasize that good models are like the real thing, but no model is exactly the same as the real thing. Ask, "What are some ways the model car is not exactly the same as a real car?" [It's smaller, has no motor, the doors don't open, the tires are metal, it doesn't have gas in it, it has no lights, it can't move under its own power, etc.] (Note: Students enjoy finding inaccuracies in models, and they are usually good at it!)

Define scale models. Tell students that although it's much smaller than a real car, this model looks like a car because someone measured every part and made each part smaller by the same factor. It is a scale model of a real car.

2-D versus 3-D models. Two-dimensional models (2-D) have height and width but are flat, like a drawing. Three-dimensional models (3-D) have height, width, and depth, like a model car.

The concept of scientific models, along with the concept of scientific evidence (presented in Session 1.4), applies to all areas of scientific endeavor and will be revisited frequently in this unit and the rest of the sequence.

SESSION 1.2 Scale Model of the Sun-Earth System

6. **Optional: You may also want to draw a two-dimensional representation of the model on a piece of paper.** Use the sharp pencil point to make a 1-mm dot (to represent Earth) and then draw a circle with a diameter of 10 cm (to represent the Sun).

7. **Modeling Earth's movement relative to the Sun.** Select a student to hold the model of the Sun. Have the class suggest what the student with the ball and the student with the pencil should do to model the movement of the Sun and Earth. [The pencil point should orbit the ball.]

Scale Model: Distance from the Earth to the Sun

1. **Distance as well as size should be in proportion.** If students have not mentioned it already, point out that in this scale model, not only should the sizes of the Sun and Earth be in the right proportion (meaning that they are the correct sizes compared to each other), but the distance between the Sun and Earth should also be in proper proportion.

2. **Have students estimate the distance between the Sun and Earth in the model.** Call on students to make guesses about the proper distance between the pencil point and the 10-cm ball to make it an accurate scale model.

3. **Model the correct distance.** Tell students that the Sun is about 150,000,000 km away from Earth. In this scale model, that distance would be represented by 12 meters, or 40 feet. Separate the students holding the model Sun and Earth by this distance. This can be done in the classroom or in a location outside of the classroom. Remind the class that Earth's orbit would take the model Earth in a big circle around the model Sun. (Don't do this unless you have the time and space available.)

One teacher said, "The students were fascinated to see the great distance between the model Sun and our pencil-point Earth. They struggled to wrap their minds around this and then transfer their new knowledge to the adapted Sun-Earth drawing."

TEACHER CONSIDERATIONS

One teacher had a few students guess at Earth's scaled distance from the model Sun by having them stand where they thought the model Earth should be. The class then voted who they thought was most accurate, before the teacher revealed where Earth would be located in this model.

SESSION 1.2 Scale Model of the Sun-Earth System

4. **The Sun puts out matter and energy.** Ask the class what the Sun puts out. [Light and heat.] Explain that in addition to energy, such as light and heat, the Sun also puts out matter in the form of charged particles. (Let the class know that they'll be learning more about this in a future session.) Have students imagine the energy and matter coming from the Sun spreading out in all directions. Ask them to describe what fraction of the Sun's output hits Earth, compared to the amount that flies off into space in other directions. Return to the classroom (if the class has moved outside) and post on the concept wall, under Key Space Science Concepts:

   The Sun is about 150,000,000 km away from Earth.

   Only a tiny amount of all the matter and energy that the Sun puts out comes toward Earth.

5. **Consider distances to other stars.** Invite students to imagine the vast distance separating Earth from other stars, given that our own star, the Sun, is so far away. In the scale of this model, the nearest star would be about 3,200 km (or about 2,000 miles) away from the pencil tip that represents Earth!

6. **Discuss how to show the scale and size of the Sun-Earth system in a drawing.** Show the third page of the Pre-unit 1 Questionnaire transparency. Have students discuss how they might now show the scale and size of the Sun-Earth system in a drawing. Explain to the class that while it may be impossible to draw the system to scale on an 8½" x 11" sheet of paper, using labels might help to communicate proper sizes and distances in a model drawing.

7. **Optional: Have your students make a scale drawing of the Sun-Earth system for use as an embedded assessment.** Refer to the Assessment Opportunity on page 85 for more details.

TEACHER CONSIDERATIONS

QUESTIONNAIRE CONNECTION
Use the questionnaire transparency to review Question #6 with your students. Refer to the scale-model activity your students have just completed and discuss with them how they would accurately represent the Sun-Earth system in a drawing, or 2-D model. You may also choose to have your students draw the Sun-Earth system as a homework assignment.

ASSESSMENT OPPORTUNITY
EMBEDDED ASSESSMENT: SCALE DRAWING OF THE SUN-EARTH SYSTEM
To assess your students' understanding of the concepts presented in this session, have them make scale drawings of the Sun-Earth system. Emphasize that students should think about what they've learned and try to represent the size and scale of the Sun-Earth system as accurately as they can. Student drawings can be used as an embedded assessment. See the scoring guide on page 85 in the Assessment section.

SESSION 1.3
Energy from the Sun

Overview

In Session 1.1, the news flashes and Information Sheets exposed students to the terms *X-ray* and *gamma ray.* In this session, these wavelengths of energy—as well as the others that comprise the electromagnetic spectrum—are systematically presented to students. A vivid display of the component colors of visible light first captures students' interest, and then the spectrum is extended beyond the visible with icons that are familiar everyday objects. The session discussion reveals that not only does the Sun radiate the full electromagnetic spectrum, but it also puts out *solar wind*—a constant stream of energetic particles. The understanding that students gain in this session of what is emitted on a "normal" day for the Sun prepares them for future sessions when they will learn about solar storms on the Sun and their effects on Earth. During this session, the key concepts that will be added to the classroom concept wall are:

- *The visible light our eyes can detect is part of a larger spectrum of electromagnetic energy.*
- *The Sun radiates the full spectrum of electromagnetic energy.*
- *The solar wind is a constant stream of charged particles that the Sun puts out.*

Energy from the Sun	Estimated Time
Introducing the Visible Spectrum	15 minutes
Beyond the Visible Spectrum	20 minutes
The Solar Wind	10 minutes
Total	**45 minutes**

Unit Goals

The Sun is a star, and a main source of energy for Earth.

The Sun gives off the full spectrum of electromagnetic energies, as well as solar particles.

The Sun's energy and matter output varies and is not constant.

Earth has protective shields located at various altitudes that help to block much of the Sun's harmful output from reaching Earth's surface.

Safety is a concern — without Earth's shields to protect us, some of the Sun's energies can be harmful.

What You Need

For the class:
- ❏ (optional) computer with large-screen monitor or LCD projector
- ❏ (optional) Space Science Sequence CD-ROM
- ❏ prepared key concept sheets from the copymaster packet or CD-ROM file
- ❏ 1 large sheet of *white* butcher paper (about 6 feet long)
- ❏ a bold, black marker (for writing on the butcher paper)
- ❏ an overhead projector
- ❏ 2 file folders
- ❏ 1 sheet of diffraction grating (about 10 cm x 10 cm, or 4" x 4") (when not in use, keep diffraction grating between the protective sheets it came with)
- ❏ tape
- ❏ 1 copy of each icon representing regions of the electromagnetic spectrum (radio, microwave, infrared, ultraviolet, X-ray, and gamma ray) from the copymaster packet or CD-ROM file

TEACHER CONSIDERATIONS

TEACHING NOTES

The timing for this session can be tight. If this is the case, move the discussion of the solar wind to the beginning of Session 1.4. Or, introduce the solar wind briefly in this session and discuss it in more detail at the beginning of Session 1.4.

One teacher said, "This lesson was the coolest lesson I have ever taught. It was the students' interest that drove it!"

Key Vocabulary

Scientific Inquiry Vocabulary

Control
Evidence
Model
Observation
Prediction
Scale
Scale model
Scientific explanation

Space Science Vocabulary

Coronal mass ejection (CME)
Electromagnetic (EM) energy
Magnetosphere
Matter
Particle
Shield
Solar flare
Solar particle
Solar wind
Spectrum
Star
Ultraviolet (UV)

SESSION 1.3 Energy from the Sun

❑ transparency of the Pre-unit 1 Questionnaire from Session 1.2
❑ (optional) several sets of colored pencils or markers
❑ (optional) electromagnetic spectrum discussion props (see Getting Ready)

For each student:
❑ 1 copy of the Solar Output Data Sheet student sheet (one page) from the copymaster packet or CD-ROM file

Getting Ready

1. **Prepare the key concept sheets.** Make a copy of each key concept and have them ready to post onto the classroom concept wall during the session.

2. **Choose a wall on which to project the spectrum.** Post the large sheet of white butcher paper on the wall and set up the overhead projector so that its light shines onto the paper.

3. **Use file folders to project only a single slit of light.** Place the two file folders side-by-side on the base plate of the projector so each folder covers one half of the projector plate. Position the file folders so there is a gap of about 2 cm (or about ¾") between them. The projector should now project a thin *vertical* slit of white light onto the paper.

4. **Tape the diffraction-grating sheet in place over the upper lens of the projector.** Use tape on the top edge of the grating only, so that it can be flipped on and off of the upper lens of the projector as necessary. With the grating in place over the lens, two spectra should appear on either side of the projected slit of white light. If this is not the case, rotate the grating by 90° until you see the two spectra and then tape the grating in place on the lens.

5. **Center one of the spectra.** Practice turning the projector slightly to the right so that one of the spectra (the one to the *left* of the slit of light as you face the butcher paper) is centered in the middle. (The centered spectrum should have red on its left and violet on its right.)

Handle the diffraction grating by its edges only—avoid touching or leaving fingerprints on the central part of the grating. When you are not using the diffraction grating, keep it between the protective sheets it came with.

TEACHER CONSIDERATIONS

TEACHING NOTES

The key concepts can be posted in many different ways. If you don't want to use sentence sheets, here are some alternatives:

- Write the key concepts out on sentence strips.
- Write the key concepts out before class on a posted piece of butcher paper. Cover each concept with a strip of butcher paper and reveal each one as it is brought up in the class discussion.

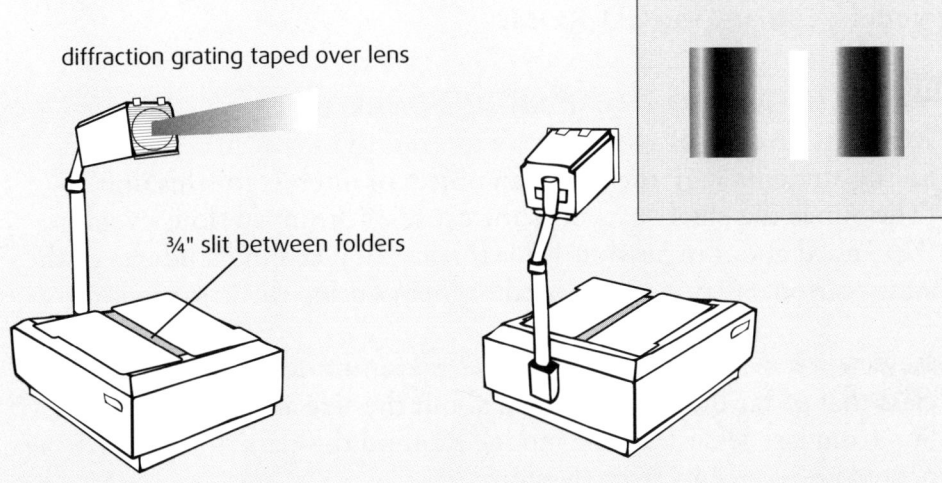

If you are teaching this activity multiple times, there is a way to use only one sheet of butcher paper. After the first time you present the activity, there will be labels written on the paper for the different types of electromagnetic radiation. For subsequent presentations, cover the labels with blank paper and then remove the paper to reveal the labels as you go.

SESSION 1.3 Energy from the Sun

6. **Optional: Gather props for the class discussion.** Read through the Beyond the Visible Spectrum discussion (pages 182–186) and gather together items that can be used to illustrate the various energies in the electromagnetic spectrum.

7. **Optional: Set up the computer and projector if you plan to use the Space Science Sequence CD-ROM.** The CD-ROM contains a useful and informative visual about the electromagnetic spectrum. Preview the Electromagnetic Spectrum activity and decide whether or not you would like to use the CD-ROM.

GO! Introducing the Visible Spectrum

1. **Remind the class of why they're interested in the Sun.** Begin by asking students why the Sun is an object of interest for this unit. [The Sun is the chief suspect as the cause of the mysterious events they heard about in Session 1.1.] (If necessary, remind students of the strange events that were revealed to them during the first session.)

2. **Review what students have learned so far about the Sun.** Tell the class that so far they have learned about the size and distance of the Sun from last session's scale model. Remind the class of the following posted key concepts from Session 1.2:

   The Sun is a star.

   The Sun is about 150,000,000 km away from Earth.

   Only a tiny amount of all the matter and energy that the Sun puts out comes toward Earth.

3. **Learning more about the energy that comes from the Sun.** Tell students that today they will learn about the kinds of energy that come from the Sun. They will begin by learning about the electromagnetic spectrum. Explain to students that what they learn today will help them better understand stars and one star in particular—our Sun.

4. **Turn on the overhead projector *without* the diffraction grating over the lens.** Darken the room, then flip the diffraction grating off the lens and turn on the overhead projector. Explain to the class that you have placed two file folders on the projector in order to create a narrow slit of light they can see on the posted sheet of butcher paper.

TEACHER CONSIDERATIONS

CD-ROM NOTES

The Electromagnetic Spectrum. This interactive is a resource guide for the electromagnetic (EM) spectrum, which can be used by both teachers and students. After the introduction screen, a slider can be used to select a specific wavelength range of the EM spectrum. The text box below the slider reveals additional information about that particular EM wavelength. Available information for each wavelength range includes: a sub-spectrum, a list of natural occurrences and common uses, a photo of the Sun, and a cool fact. Click SEE SCIENCE IN SPACE in the lower right-hand corner to display a page on how NASA uses the selected wavelength range to observe phenomena in space. A waveform is provided above the EM spectrum to show wavelength in metric units. To enlarge the interactive to full screen, click CONTROL F for Windows and APPLE F for Macs. Click ESC to exit this mode. You can close the program any time, just as you would close a window on your desktop.

If time allows, we suggest you show students the video, *Infrared: More Than Your Eyes Can See*, which features a scientist named Michelle Thaller. This eight-minute video is very involving.

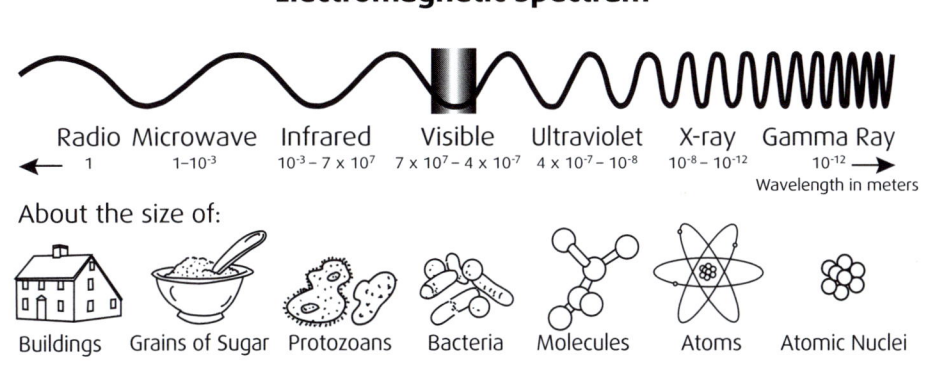

SESSION 1.3 Energy from the Sun

5. **Explain a diffraction grating.** Tell the class that a *diffraction grating* is a special piece of scored plastic that can be used to more carefully analyze light. Ask the class to observe what happens when the light from the projector passes through a diffraction grating. Flip the diffraction grating in place over the lens of the projector. Ask your students to describe what they see. [Lots of different colors.]

6. **The diffraction grating separates white light into a spectrum of colors.** Ask students where they think the different colors come from. Give them a few minutes to offer their suggestions and then reveal to the class that white light is made up of all these colors together. Explain that the diffraction grating separates the white light from the projector into a *spectrum* of colors. Explain that if light from the Sun also passed through a diffraction grating, it would create a spectrum similar to the one that they are looking at.

7. **Center one of the spectra.** Turn the projector slightly to the right so that one of the spectra (the one to the *left* of the slit of light as you face the butcher paper) is centered in the middle. Ask the class which colors are at either end of the spectrum. [Red is on the left and violet is on the right.]

8. **Have students list the colors of the spectrum in order from left to right.** As students list the colors out loud, write the names of the colors on the butcher paper above their corresponding bands in the spectrum. (Note: There are many colors in the spectrum and many names that can be used to describe them. It's not important how many colors are used or how they are named, as long as you have red at one end, purple or violet at the other end, and a fair selection of colors in between.)

9. **Explain the relationship between a color's position in the spectrum and its energy.** Tell students that light is made up of "bundles of energy" called *photons*. Each color of light is made up of photons with different energies, and these energies determine a color's place in the spectrum. The energy carried by a photon of violet light is about twice as much as the energy of a photon of red light. Ask the class questions to gauge their understanding of this: "Which color has photons with more energy: yellow or blue?" [Blue.] "Green or violet?" [Violet.] "Red or green?" [Green.]

10. **Mark energy levels of colors at the ends of the spectrum.** Use the black marker and write "Higher Energy" on the upper-right side of the paper above the violet band of light and "Lower Energy" on the upper-left side of the paper above the red band of light.

TEACHER CONSIDERATIONS

SESSION 1.3 Energy from the Sun

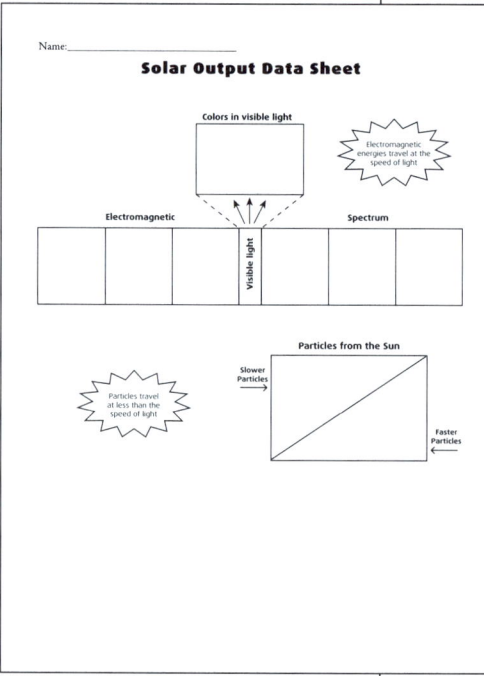

11. **Have the class record the colors of the visible spectrum on their student sheets.** Pass out a copy of the Solar Output Data Sheet to each student. (If you have colored pencils or markers for the class, pass these out as well.) Explain that they have been looking at the colors in the *visible spectrum*. Have students label (or color in) the colors of the visible spectrum in the box labeled Colors in visible light.

12. **Moving beyond the visible spectrum.** Continue to project the visible spectrum as you introduce the rest of the electromagnetic spectrum.

Beyond the Visible Spectrum

1. **Explain that there is more to the electromagnetic spectrum than the visible spectrum.** Tell the class that there is more to the electromagnetic spectrum than what they currently see. In 1800, Sir Frederick William Herschel split up sunlight using a glass prism instead of a diffraction grating. He then placed thermometers on various colors of the visible spectrum and also on either side of the spectrum to measure the heating effects of different colors of light in the spectrum. Surprisingly, he found that even the thermometer to the left of red, beyond the visible spectrum, heated up. Herschel had discovered a type of energy not visible to humans called infrared, or IR. (Later, scientists learned that some animals are able to perceive this energy.)

2. **Discussing infrared.** Ask students what they've heard about infrared. [Infrared night-vision instruments allow us to see in the dark, detecting the IR radiation given off by warm objects. The beam of the TV remote is infrared, invisible to our eyes.] If possible, demonstrate infrared in the classroom by using a real remote control to operate a TV or VCR.

3. **Mark infrared on the spectrum.** Write "infrared" on the butcher paper immediately to the left of the visible spectrum and tape the icon for the TV remote above it. Have students record "infrared" on their student data sheets.

4. **Have the class compare the energy of infrared and visible light.** Ask, "Do you think infrared has more or less energy than red light?" [Less. As you move to the left on the spectrum, the energy carried by the photons is lower.]

5. **Introduce UV.** Tell them that there is also a type of energy beyond violet. Point to the area on the butcher paper to the right of the violet end of the spectrum. It's invisible to humans, but bees can see it. Ask what it's called. [Ultraviolet or UV light.]

TEACHER CONSIDERATIONS

TEACHING NOTES
To learn about how diffraction gratings work and how they differ from prisms, see page 36 in the Background Information for Teachers.

SCIENCE NOTES
The longer wavelengths of infrared from the Sun can only be detected from mountain tops (above most of the water vapor) and from space. Although it heats the atmosphere as it's absorbed, most solar infrared doesn't reach Earth's surface. However, some shorter wavelength infrared does reach Earth's surface, which was where Herschel first detected it. See the Background Information for Teachers section, page 34 for more on infrared. For an excellent discussion and illustration of which regions of the electromagnetic spectrum reach Earth's surface, see the Amazing Space website at: http://amazing-space.stsci.edu/resources/explorations/groundup/lesson/basics/g17b/index.php

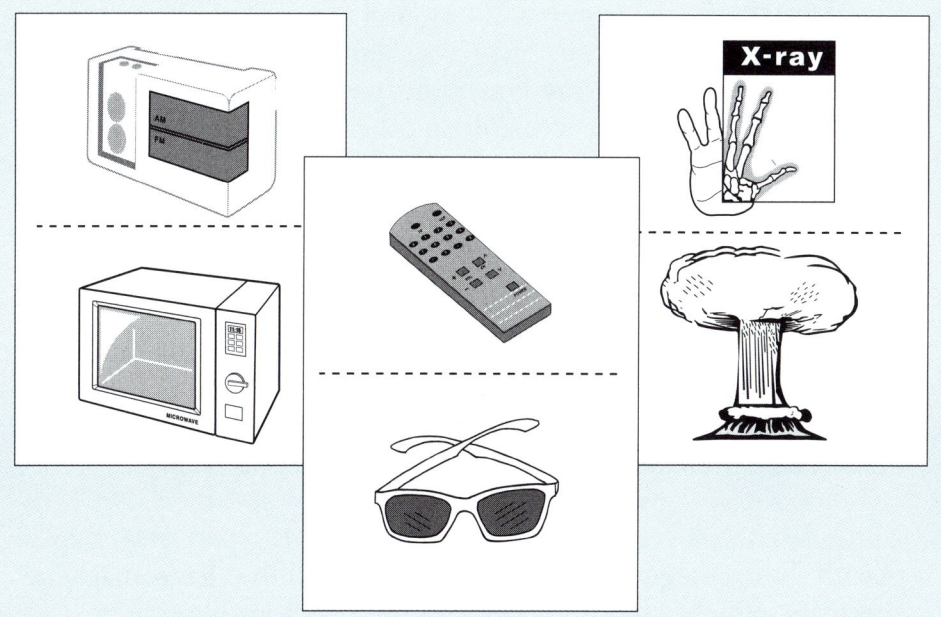

SESSION 1.3 Energy from the Sun

6. **Compare the energy of ultraviolet and visible light.** Ask, "Judging by where ultraviolet energy is located on this spectrum, do you think it has more or less energy than visible light?" [More.]

7. **Discussing UV.** Ask what students have heard about ultraviolet. [It can cause sunburn; it can also damage our eyes. Sunscreen and some sunglasses partially block it—but it is **NOT** safe to look directly at the Sun.] If possible, hold up a real pair of sunglasses to highlight the ultraviolet part of the spectrum.

8. **Mark UV on the spectrum.** Write "ultraviolet" on the butcher paper immediately to the right of the visible spectrum and tape the icon of the sunglasses above it. Have students add "ultraviolet" to their sheets.

9. **Discussing other energies.** Ask the class if they know of other invisible energies that are lower than infrared and/or higher than ultraviolet. [Radio, microwave, X-ray, and gamma ray.] Ask students to share any information they may have about these energies. If possible, hold up a real X-ray film, radio, pager, cell phone and/or any other props you may have to highlight other parts of the spectrum.

10. **Fill in the remaining parts of the spectrum.** Add the labels and icons for radio, microwave, X-rays, and gamma rays to the butcher paper and have students record them on their data sheets.

11. **Mark energy on the butcher paper.** Write across the top (above the labels and icons) "Low Energy" to the far left and "High Energy" to the far right. Draw an arrow running from left to right between them. Have students add this to their data sheets as well.

12. **All these energies make up the electromagnetic spectrum.** Explain that all these energies together are what we call the electromagnetic spectrum, which is like an extended rainbow of visible and invisible energies. Label the class chart on butcher paper: The Electromagnetic Spectrum. Post on the concept wall, under Key Space Science Concepts:

The visible light our eyes can detect is part of a larger spectrum of electromagnetic energy.

One teacher said, "There were lively discussions about the different energy levels and the effects on us and life on Earth. The use of pictures for each level helped by creating a mental image. I think they understood the difference between the energy levels of the spectrum and the matter of solar winds and solar particles."

TEACHER CONSIDERATIONS

SCIENCE NOTES
Physicists measure the energy of electromagnetic photons in electron volts (or eV), as well as other units. However, for the purposes of this unit, students need only know that there is a huge range of energies in the electromagnetic spectrum, going from low (radio) to very high (gamma rays).

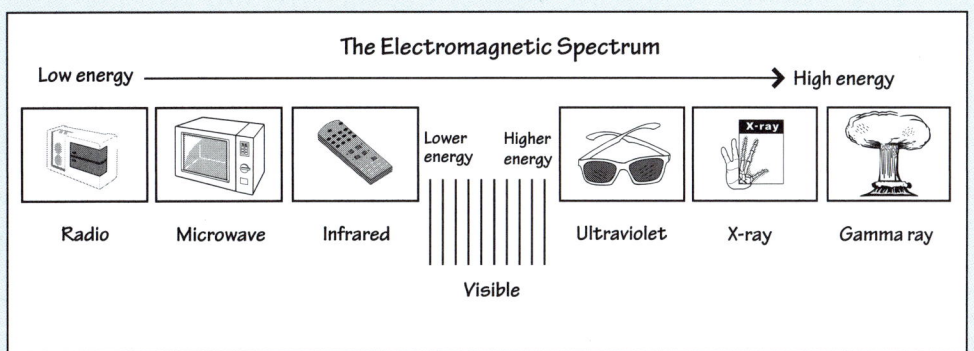

SESSION 1.3 Energy from the Sun

13. The Sun radiates the full spectrum of electromagnetic energy. Point out that visible light is only a small part of the entire electromagnetic spectrum. Explain that the Sun gives off energy from all parts of the spectrum. Post on the concept wall, under Key Space Science Concepts:

The Sun radiates the full spectrum of electromagnetic energy.

14. Certain energies can be harmful. If students have not yet mentioned that certain energies can be harmful, be sure to bring it up at this point. Mention that at times, microwave, UV, X-ray, and gamma-ray energies can be harmful to living things. In later sessions, students will learn some of the ways in which humans are shielded from these harmful energies. Also mention, though, that these energies can also be helpful. UV, for example, can be beneficial to plants in limited amounts, and X-rays are useful for medical purposes.

The Solar Wind

1. Besides electromagnetic energy, particles also come from the Sun. Tell the class that the Sun doesn't just put out energy from the electromagnetic spectrum. The Sun also puts out a constant stream of charged particles called the *solar wind*.

2. The difference between particles and electromagnetic energies. Explain that particles are different from the electromagnetic energies the class has been studying. Particles are matter, which means they have size and mass. Say that protons and electrons are both examples of particles. Explain that the solar wind is a constant flow of charged particles that streams outward from the Sun. Post on the concept wall, under Key Space Science Concepts:

The solar wind is a constant stream of charged particles that the Sun puts out.

3. Particles and electromagnetic energies travel at different speeds. Explain that another difference between particles and electromagnetic energies is the speed at which they travel through space. All electromagnetic energy travels at the speed of light. Nothing else travels faster than the speed of light, so solar-wind particles move more slowly—much more slowly—than electromagnetic energies.

TEACHER CONSIDERATIONS

TEACHING NOTES

Be sure to save the Electromagnetic Spectrum chart—it will be used frequently in this unit.

The topic of solar wind may be difficult for your students to grasp initially. The first part of Session 1.4 reviews and builds upon what is learned about the solar wind in this session, so it's okay to briefly introduce the idea to students now.

SESSION 1.3 Energy from the Sun

One teacher said, "Several students were surprised to learn that everything that the Sun emits doesn't travel at the speed of light."

4. **Solar wind particles move slower than electromagnetic energies.** Explain that even though most particles from the Sun travel hundreds of kilometers a second, that is still much slower than the speed of light. (Light travels at 300,000 kilometers per second.) So, it takes the solar wind a few days to get from the Sun to Earth, while electromagnetic energy from the Sun reaches us in just about eight minutes.

5. **Have students add the solar wind to their student sheets.** Point out the box labeled Particles from the Sun on the Solar Output Data Sheet. Have students write "The Solar Wind" in the upper-left half of the box labeled Slower Particles. Explain that in the next session they will learn about faster particles, which come from the Sun.

6. **Review of key concepts introduced today.** Tell the class that they have learned three key concepts about the Sun today that are especially important for them to remember as they continue their study of solar science. Refer them to this session's key concepts, posted on the concept wall:

   The visible light our eyes can detect is part of a larger spectrum of electromagnetic energy.

   The Sun radiates the full spectrum of electromagnetic energy.

   The solar wind is a constant stream of charged particles that the Sun puts out.

7. **Review Question #1 of the Pre-unit 1 Questionnaire with students.** Show the transparency of the questionnaire. Have students discuss how they might answer Question #1 now.

8. **The Solar Output Data Sheet will be used during the next session.** Students will need to refer to their student sheets for Session 1.4. Decide whether you would like to collect the student sheets or have students hold on to them for use next time.

9. **Save the sheet of butcher paper with the electromagnetic spectrum on it for use in Sessions 1.4 and 1.5.** The class will need to refer to the Electromagnetic Spectrum chart in future sessions.

TEACHER CONSIDERATIONS

TEACHING NOTES
Although it's beyond the scope of this unit, there's actually a whole range of solar particles, from low to high energy. Some of the highest energy particles do travel near the speed of light. For more information, see the Background Information for Teachers section, page 33.

QUESTIONNAIRE CONNECTION
Review Question #1 on the questionnaire with your students. Students should notice that three of the four choices are EM energies, and that only **C**, particles, is not an EM energy.

ASSESSMENT OPPORTUNITY
QUICK CHECK FOR UNDERSTANDING: THE SUN'S OUTPUT
At the end of this session, students should understand that the Sun's output includes solar particles as well as the full spectrum of electromagnetic energies.

SESSION 1.4

A Stormy Sun: Revisiting the Mystery

Overview

During a solar storm, huge gusts of electromagnetic energies and solar energetic particles are ejected from the Sun into space by solar flares and coronal mass ejections (CMEs). If the output of a solar storm reaches Earth, events such as those introduced by the news flashes in Session 1.1 can occur. The introduction of solar storms, solar flares, and CMEs in this session teaches students that the energy that reaches Earth from the Sun is not constant, but variable. In this session, students learn that the Sun has periods of intense activity when solar flares and coronal mass ejections can occur. Drawing from their knowledge of solar science, student teams discuss and write answers to questions about the mystery events in preparation for their roles as "expert witnesses" in a trial implicating the Sun as the culprit in the mystery. The students' "testimony" brings out the intriguing match between the dates of the disruptions on Earth and the earlier occurrence of a huge solar storm. During this session, the key concepts that will be added to the classroom concept wall are:

- *Solar flares and coronal mass ejections (CMEs) occur during solar storms, when the Sun is active.*
- *A solar flare releases large amounts of electromagnetic energies and solar particles into space.*
- *A CME ejects particles and material from the Sun's corona at high speeds into space.*
- *Particles released by solar-storm events—such as solar flares and CMEs—travel much more quickly than particles in the solar wind.*
- *The amount of particles and energy put out by the Sun is not constant.*
- *Scientific explanations are based on evidence gathered from observations and investigations.*

Unit Goals

The Sun is a star, and a main source of energy for Earth.

The Sun gives off the full spectrum of electromagnetic energies, as well as solar particles.

The Sun's energy and matter output varies and is not constant.

Earth has protective shields located at various altitudes that help to block much of the Sun's harmful output from reaching Earth's surface.

Safety is a concern — without Earth's shields to protect us, some of the Sun's energies can be harmful.

A Stormy Sun: Revisiting the Mystery	Estimated Time
An Active Sun: Solar Flares and Coronal Mass Ejections	10 minutes
Expert-Witness Testimony Preparation	20 minutes
Class Role Play: The Sun on Trial	15 minutes
Total	**45 minutes**

What You Need

For the class:
- ❑ overhead projector or computer with large-screen monitor or LCD projector
- ❑ prepared key concept sheets from the copymaster packet or CD-ROM file

TEACHER CONSIDERATIONS

TEACHING NOTES

The key concepts can be posted in many different ways. If you don't want to use sentence sheets, here are some alternatives:

- Write the key concepts out on sentence strips.
- Write the key concepts out before class on a posted piece of butcher paper. Cover each concept with a strip of butcher paper and reveal each one as it is brought up in the class discussion.

In later sessions, students will gather more evidence and confirm that the Sun did, indeed, cause the mysterious events. Questions about why some of the Sun's effects are felt only above the surface of Earth and not on the ground are dealt with in the next session.

Key Vocabulary

Scientific Inquiry Vocabulary

Control
Evidence
Model
Observation
Prediction
Scale
Scale model
Scientific explanation

Space Science Vocabulary

Coronal mass ejection (CME)
Electromagnetic (EM) energy
Magnetosphere
Matter
Particle
Shield
Solar flare
Solar particle
Solar wind
Spectrum
Star
Ultraviolet (UV)

SESSION 1.4 A Stormy Sun: Revisiting the Mystery

- the Electromagnetic Spectrum chart from Session 1.3
- week-long calendar of mysterious events from Session 1.1
- 1 copy of the Particles from the Sun mini-poster (one page) from the copymaster packet or CD-ROM file
- transparency of the Quiet Sun/Active Sun Graph (one page) from the transparency packet or CD-ROM file
- (optional) 1 copy of the Expert Witness Role-play Script (one page) from the copymaster packet or CD-ROM file
- transparencies of the Pre-unit 1 Questionnaire from Session 1.2 (three pages)

For each student:
- 1 copy of the Expert Witness Questions student sheet (two pages) from the copymaster packet or CD-ROM file
- completed Solar Output Data Sheet student sheet from Session 1.3
- completed Research Notes student sheet from Session 1.1

Getting Ready

1. **Prepare the key concept sheets.** Make a copy of each key concept and have them ready to post onto the classroom concept wall during the session.

2. **Post the Electromagnetic Spectrum chart and the week-long calendar of mysterious events on a classroom wall.** Choose a location that every student will be able to see.

3. **Make a copy of the Particles from the Sun display sheet.** This will be added to the Electromagnetic Spectrum chart during a class discussion.

4. **Decide how you will divide the class into small groups for the Expert-Witness Testimony Preparation activity.**

5. **Prepare the student sheets for this session.** Make a copy of the Expert Witness Questions student sheets for each student. If you collected the Research Notes and Solar Output Data Sheet student sheets from previous sessions, prepare to hand them out. Otherwise, ask the students to have these completed student sheets ready to use.

6. **Decide who you would like to play the role on the lawyer during the mock trial.** You may choose a student, or play the role yourself. Look through the Expert Witness Role-play Script. If you decide to assign the role to a student, make a copy of the script.

7. **Decide whether you will use the Expert Witness Questions student sheets as an assessment.** See page 201 for more information.

One teacher said, "I selected a panel of expert witnesses to have more students involved. I added some additional questions as a 'cross-examination' to address some of their ideas from the first day of the unit. For example, some students had suggested that an earthquake might have occurred. We added this as one of the questions for the panel to address, and they were able to rule it out."

TEACHER CONSIDERATIONS

SESSION 1.4 A Stormy Sun: Revisiting the Mystery

GO! An Active Sun: Solar Flares and Coronal Mass Ejections

1. **Pass out the Solar Output Data Sheet and the Research Notes student sheets collected from previous sessions.** If you opted to have the students hold onto their sheets, ask them to take them out now.

2. **Review what comes from the Sun.** Remind the class that last time, they learned about what the Sun puts out. Ask students what comes from the Sun. [All energies in the electromagnetic spectrum come from the Sun. Charged particles making up the solar wind also flow outward from the Sun.] If necessary, refer the class to their Solar Output Data Sheets as well as the posted sheet of butcher paper with information about the EM spectrum.

3. **The Sun has two states of activity.** Say that the Sun experiences periods of high activity as well as periods of low activity. Explain that scientists describe the Sun as *active* when it's in a state of high activity and *quiet* when it's in a state of low activity.

4. **Solar storms occur when the Sun is active.** Explain that when the Sun is in an active state, solar storms can occur. During a solar storm, there are energetic events such as solar flares and coronal mass ejections (or CMEs). Post on the concept wall, under Key Space Science Concepts:

   Solar flares and coronal mass ejections (CMEs) occur during solar storms, when the Sun is active.

5. **Reviewing solar flares from Session 1.1.** Remind students that at the beginning of the unit, they researched suspects that might be responsible for several mysterious events. One of those suspects was solar-surface activity, which referred to solar flares. Have students refer to their Research Notes student sheets, then ask what a solar flare is. [A solar flare is an extremely bright area on the Sun's surface that lasts for only a few hours. A flare is more likely to occur during periods of high sunspot activity.] Explain that a solar flare erupts from a part of the Sun's upper atmosphere, or *corona*. A solar flare releases large amounts of electromagnetic energy and solar particles into space. Post on the concept wall, under Key Space Science Concepts:

   A solar flare releases large amounts of electromagnetic energies and solar particles into space.

TEACHER CONSIDERATIONS

SCIENCE NOTES
Please see page 32 of the Background Information for Teachers section for more information on these phenomena and their connection to solar storms.

SESSION 1.4 A Stormy Sun: Revisiting the Mystery

6. **Discussing coronal mass ejections, or CMEs.** Say that during a solar storm, coronal mass ejections, or CMEs, may also occur. Like a solar flare, a CME releases gusts of solar particles. However, CMEs are much larger events than solar flares, ejecting material from the Sun's corona at high speeds into space. Post on the concept wall, under Key Space Science Concepts:

   A CME ejects particles and material from the Sun's corona at high speeds into space.

7. **Solar-storm particles move much faster than solar-wind particles.** Explain that the large amount of particles sent out during a solar-storm event, such as a solar flare or a CME, travels much more quickly than particles in the solar wind. Remind the class of the posted key concept from Session 1.3:

   The solar wind is a constant stream of charged particles that the Sun puts out.

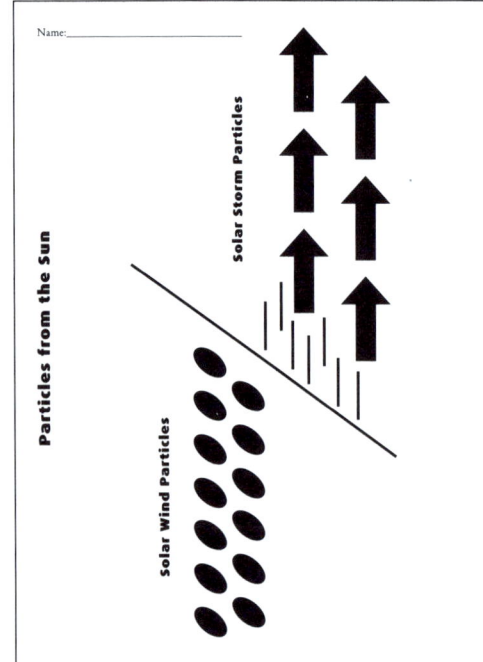

Post the Particles from the Sun display sheet on the Electromagnetic Spectrum chart. On their Solar Output Data Sheets, have students write "A Solar Storm" in the lower right-hand corner of the Particles from the Sun box labeled Faster Particles. Post on the concept wall, under Key Space Science Concepts:

Particles released by solar-storm events—such as solar flares and CMEs—travel much more quickly than particles in the solar wind.

8. **Particles are made of matter.** Emphasize that the particles released by solar flares and CMEs, like the regular solar wind, are made of matter. Remind the class that matter travels more slowly than electromagnetic energies such as X-rays and gamma rays.

TEACHER CONSIDERATIONS

SCIENCE NOTES
When a CME gust erupts from the Sun and plows into the solar wind, it creates a shock wave, which in turn accelerates solar particles to a very high speed. If these head toward Earth, they endanger astronauts and spacewalkers almost immediately. The gust of CME matter behind the particles can take 2–3 days to arrive and can cause geomagnetic interference on Earth. For more information on high-speed particles (called SEPs), CMEs, solar flares, and the solar wind, see the Background Information for Teachers section, page 32.

TEACHING NOTES
For the purposes of this unit, students should understand that particles (matter) travel more slowly than electromagnetic energy. You may also want to emphasize to them that not every solar-storm event affects Earth—only those that happen to come in Earth's direction.

SESSION 1.4 A Stormy Sun: Revisiting the Mystery

9. **The Sun's output is not constant.** Have the class look at the posted Electromagnetic Spectrum from Session 1.3 as well as the posted Particles from the Sun display sheet. Point out that both electromagnetic energies and solar particles come from the Sun all the time. Ask, "Is the amount of the Sun's energy and particle output always the same, though?" [No. During solar storms, the output of energies and particles from the Sun increases.] Post on the concept wall, under Key Space Science Concepts:

   The amount of particles and energy put out by the Sun is not constant.

   Also remind students that, as they learned earlier, only a small part of the Sun's ouput reaches Earth. Refer to the posted key concept from Session 1.2:

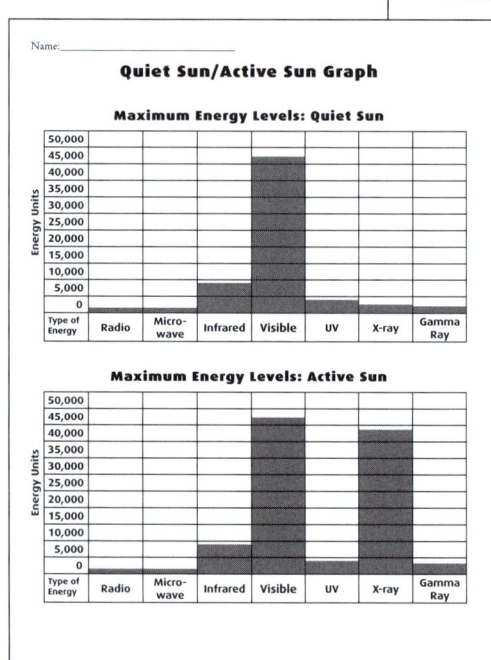

   Only a tiny amount of all the matter and energy that the Sun puts out comes toward Earth.

10. **Comparing the output between a quiet Sun and an active Sun.** Show the Quiet Sun/Active Sun Graph transparency. Explain to the class that the two graphs show electromagnetic energy from the Sun when it is quiet and when it is active. Ask students what differences in energy level they notice between the two graphs. Students should notice that there are significant increases in both X-rays and gamma rays when the Sun is active.

11. **Review Question #4 of the Pre-unit 1 Questionnaire with students.** Show the transparency of the Pre-unit 1 questionnaire. Have students discuss how they might answer Question #4, now that they have learned more about the Sun's output.

Expert-Witness Testimony Preparation

1. **Preparing to become expert witnesses in solar science.** Tell the class that with their newly gained knowledge about the Sun, they are about to play an important part in solving the case of the mysterious events they heard about in Session 1.1. Remind the class that the Sun is currently a chief suspect in the cause of the events. Ask students to imagine that the Sun is now on trial. A lawyer for the prosecution would like to ask some solar-science expert witnesses for evidence that the Sun caused the mysterious events. Explain to students that in the trial that's about to begin, they will be playing the roles of the expert witnesses!

TEACHER CONSIDERATIONS

QUESTIONNAIRE CONNECTION
Go over Question #4 on the Pre-unit 1 questionnaire with your students. Encourage them to use the information they have learned about the electromagnetic spectrum, solar storms, solar flares, and CMEs to provides reasons—or evidence—for their answers.

TEACHING NOTES
Not every student will feel comfortable, or enjoy, playing the role of an "expert witness." Be sensitive to students who are shy and look for other ways to involve them in the trial. One option would be to break the class into group witness panels that discuss and come up with consensus answers to the lawyer's questions.

SESSION 1.4   A Stormy Sun: Revisiting the Mystery

> **Expert Witness Questions**
> 1. How far from the Sun is Earth?
> 2. What mysterious events happened on Earth on April 21?
> 3. Was the Sun's output normal on April 21?
> 4. What happened on the Sun at 3:10 p.m. Pacific Daylight Time on April 21?
> 5. What is a solar flare?
> 6. If the solar flare happened at 3:10 p.m. Pacific Daylight Time, why would the events occur on Earth at 3:18 p.m. Pacific Daylight Time—eight minutes later?

2. **Review with students the resources available to them.** Tell students that to prepare for their roles as expert witnesses, they should review their Research Notes and Solar Output Data Sheet student sheets. Say that they can also refer to the following resources posted around the classroom: the weeklong calendar of mysterious events, the key concepts on the concept wall, and the Electromagnetic Spectrum class chart with the Particles from the Sun mini-poster.

3. **Working on the Expert Witness Questions student sheets in small groups.** Tell students that the Expert Witness Questions student sheets will help prepare them for questions the lawyer may ask during the trial. Say that they will be working in small groups to answer the questions on the student sheets. Emphasize that although they will be working in groups, each student is responsible for completing his or her own sheets.

4. **Divide the class into small groups.** Tell students that they will have about 20 minutes to prepare their "testimonies." Pass out the Expert Witness Questions student sheets and have groups begin their work. Let the class know when they have only five minutes left.

5. **Regain the attention of the class.** Tell the class that you will be playing the role of the lawyer, or choose a student to play the role and give him or her a copy of the Expert Witness Role-play Script student sheet.

Class Role-play: The Sun on Trial

1. **Begin the trial.** Using the Expert Witness Role-play Script below, ask the questions of various students, probing for details when necessary.

Expert Witness Role-play Script
[Likely answers from the expert witnesses are included in brackets.]

"Thank you for being our expert witnesses today in the field of solar science. Let's begin with this question: How far from the Sun is Earth?" [About 150,000,000 km away.] If necessary, remind students of the scale-model activity and key concept from Session 1.2. Assure students that scientists have evidence that this distance is accurate. (See the Background Information for Teachers section, page 29.)

"What mysterious events happened on Earth on April 21?"
[At 3:18 p.m. Pacific Daylight Time, radio transmissions were interrupted, and increased X-rays and gamma rays were detected.]

TEACHER CONSIDERATIONS

ASSESSMENT OPPORTUNITY
EMBEDDED ASSESSMENT: EXPERT-WITNESS QUESTION RESPONSES
Student responses on the Expert-Witness Questions student sheet can be used as an embedded assessment (specifically Question #s 6 and 9). See the scoring guide on page 87 in the Assessment section.

SESSION 1.4 A Stormy Sun: Revisiting the Mystery

Expert Witness Role-play Script

Name:_____

1. Thank you for being our expert witnesses today in the field of solar science. Let's begin with this question: How far from the Sun is Earth?
2. What mysterious events happened on Earth on April 21?
3. Was the Sun's output normal on April 21? Please answer yes or no.
4. What happened on the Sun at 3:10 p.m. Pacific Daylight Time on April 21?
5. Tell us, what is a solar flare?
6. Let's assume for a moment that the solar flare caused the events on Earth previously mentioned. If the solar flare happened at 3:10 p.m. Pacific Daylight Time, why would the events occur on Earth at 3:18 p.m. Pacific Daylight Time—eight minutes later?
7. Along with the big solar flare, there was a coronal mass ejection (or CME) on April 21. What is a CME and do particles from CMEs travel at the speed of light?
8. What happened on Earth on April 23?
9. How could the CME on April 21 cause the events on Earth on April 23?

"Was the Sun's output normal on April 21?" Please answer yes or no. [No.]

"What happened on the Sun at 3:10 p.m. Pacific Daylight Time on April 21?" [A large solar flare erupted at 3:10 p.m. Pacific Daylight Time.]

"Tell us, what is a solar flare?" [A solar flare is an extremely bright area on the surface of the Sun. It sends out a burst of particles and electromagnetic energies.]

"Let's assume for a moment that the solar flare caused the events on Earth previously mentioned. If the solar flare happened at 3:10 p.m. Pacific Daylight Time, why would the events occur on Earth at 3:18 p.m. Pacific Daylight Time—eight minutes later?" [The Earth is 150,000,000 kilometers away from the Sun. It takes eight minutes for the Sun's electromagnetic energies to reach Earth, even though they travel at the speed of light. (The speed of light is 300,000 kilometers per second, or 186,000 miles per second.)]

"Along with the big solar flare, there was a coronal mass ejection (or CME) on April 21. What is a CME and do particles from CMEs travel at the speed of light?" [A CME is a gust of particles and material from the Sun. The solar-storm particles associated with a CME are made of matter. They move very fast, but generally more slowly than electromagnetic energies, which travel at the speed of light. A CME gust reaches Earth two to three days after leaving the Sun.]

"What happened on Earth on April 23?" [TV broadcasts were disrupted, unusual clouds of light and color were seen in the sky, a power surge caused electrical blackouts in several cities, magnetic North shifted slightly, and cell phones stopped working.]

"How could the CME on April 21 cause the events on Earth on April 23?" [The Earth is very far—150,000,000 kilometers—from the Sun. Solar particles travel more slowly than electromagnetic energies; they take two to three days to get to Earth, which may be why many events happened two days after the large solar flare.]

2. **Thank the expert witnesses for their responses.** If necessary, review the concepts of distance and relative speeds involved. Acknowledge that it is suspicious how well the timing of the large solar flare and CME matches the timing of the disruptions on Earth. Tell the class that they'll need to learn more to be sure that the Sun did, indeed, cause these events.

TEACHER CONSIDERATIONS

SESSION 1.4 A Stormy Sun: Revisiting the Mystery

3. **Students are behaving like scientists.** Point out to students that they have just used their research notes and other information to present *scientific evidence* that the Sun may have caused the mysterious events. Post on the concept wall, under Key Scientific Inquiry Concepts:

   Scientific explanations are based on evidence gathered from observations and investigations.

4. **Scientists require scientific evidence.** Explain that scientists don't agree something is real or accurate unless they have scientific evidence. If scientists can see the evidence for themselves, test it, and confirm it, they may agree that it is scientific evidence, and that it is convincing.

5. **Gathering more scientific evidence about the Sun.** Say that in the coming sessions, the class will conduct a mock mission as well as an experiment to gather more scientific evidence about how the Sun affects those on Earth—when it is quiet *and* when it is active—and how humans can protect themselves from some of the Sun's harmful effects.

TEACHER CONSIDERATIONS

ASSESSMENT OPPORTUNITY
CRITICAL JUNCTURE: REVIEWING HARMFUL EM ENERGIES

In Session 1.5, students learn more about Earth's various protective shields. Before proceeding with the Balloon-Rocket Mission graphing activity, check to make sure your students understand which electromagnetic energies can be harmful to humans. If needed, use the Electromagnetic Spectrum chart from Session 1.3 to review this with them.

One teacher said, "The first time they had to gather evidence they were a little at a loss. In most groups there was at least one person who felt they knew what to do. As more evidence was uncovered and as more misconceptions were changed they all became more confident. When the answer finally hit some of the students it was the classic 'Aha!' moment. After the expert witness activity they really felt more confident."

SESSION 1.5

The Balloon-Rocket Mission

Overview

Students are asked to advise four people located at different altitudes (from sea level to Earth orbit) of their risks due to exposure to the Sun's energy. To learn more about solar-energy levels and shields at different altitudes, the class collects data from an imaginary balloon-rocket mission. In this session, student teams work together to graph and compare data obtained from the balloon-rocket mission. Analysis of the completed graphs show that shields at various altitudes prevent some solar energies from reaching Earth's surface. The session ends with a class discussion about the Earth's magnetosphere and its role as a shield against solar particles. During this session, the key concepts that will be added to the classroom concept wall are:

- *Shields at various altitudes prevent some solar energies from reaching Earth's surface.*
- *The magnetosphere shields Earth from solar particles at a very high altitude.*
- *The changing shape of Earth's magnetosphere during a solar storm can leave satellites and astronauts unshielded.*

The Ballon-Rocket Mission	Estimated Time
Introducing the Mission	5 minutes
Preparing to Receive and Interpret Data	10 minutes
The Balloon-Rocket Mission	15 minutes
Interpreting the Data and Identifying Shields	10 minutes
Solar Particles and the Earth's Magnetosphere	5 minutes
Total	45 minutes

What You Need

For the class:
- ❑ (optional) overhead projector or computer with large-screen monitor or LCD projector
- ❑ (optional) Space Science Sequence CD-ROM
- ❑ prepared key concept sheets from the copymaster packet or CD-ROM file
- ❑ the Electromagnetic Spectrum chart with Particles from the Sun display sheet attached (from Sessions 1.3 and 1.4)
- ❑ one transparency each of the following Completed Balloon-Rocket Mission Graphing Sheets: Radio, Microwave, and Visible Light from the transparency packet or CD-ROM file (see Getting Ready)

Unit Goals

The Sun is a star, and a main source of energy for Earth.

The Sun gives off the full spectrum of electromagnetic energies, as well as solar particles.

The Sun's energy and matter output varies and is not constant.

Earth has protective shields located at various altitudes that help to block much of the Sun's harmful output from reaching Earth's surface.

Safety is a concern — without Earth's shields to protect us, some of the Sun's energies can be harmful.

TEACHER CONSIDERATIONS

TEACHING NOTES

The balloon-rocket mission requires that students be proficient at graphing. If necessary, review basic graphing skills with your students before starting this session or plan to leave plenty of additional time to complete this activity.

Key Vocabulary

Scientific Inquiry Vocabulary

Control
Evidence
Model
Observation
Prediction
Scale
Scale model
Scientific explanation

Space Science Vocabulary

Coronal mass ejection (CME)
Electromagnetic (EM) energy
Magnetosphere
Matter
Particle
Shield
Solar flare
Solar particle
Solar wind
Spectrum
Star
Ultraviolet (UV)

SESSION 1.5 The Balloon-Rocket Mission

- ❑ 1 copy each of the following Completed Balloon-Rocket Mission Graphing Sheets: Radio, Microwave, Visible Light, Infrared, Ultraviolet, X-ray, Gamma Ray, and Solar Particles from the copymaster packet or CD-ROM file (to post onto the EM chart)
- ❑ transparency of the Data for the Balloon-Rocket Mission from the transparency packet or CD-ROM file
- ❑ 1 sheet of paper
- ❑ transparency of the Completed Balloon-Rocket Mission Graphing Sheet: Solar Particles from the transparency packet or CD-ROM file
- ❑ transparency of the Extension Graph of Solar Particles from the transparency packet or CD-ROM file
- ❑ 1 copy of the Extension Graph of Solar Particles from the copymaster packet or CD-ROM file (to post onto the EM chart)
- ❑ transparency of the Earth's Magnetosphere from the transparency packet or CD-ROM file
- ❑ week-long calendar of mysterious events from Session 1.1

For each team of 4 students:
- ❑ 1 copy each of the following Student Balloon-Rocket Mission Graphing Sheets: Infrared, Ultraviolet, X-ray, and Gamma Ray from the copymaster packet or CD-ROM file
- ❑ (optional) 1 copy each of the following Student Balloon-Rocket Mission Graphing Sheets: Radio, Microwave, Visible Light, and Solar Particles from the copymaster packet or CD-ROM file

Getting Ready

1. **Arrange for the appropriate projector format to display images to the class.** Decide whether you will be using the overheads or the CD-ROM. Set up an overhead projector or a computer with a large-screen monitor or LCD projector.

2. **Prepare the key concept sheets.** Make a copy of each key concept and have them ready to post onto the classroom concept wall during the session.

3. **Post the Electromagnetic Spectrum chart and week-long calendar of mysterious events.**

4. **Decide how you will divide the class into teams of four.** Students will be working in teams to graph the balloon-rocket mission data.

TEACHER CONSIDERATIONS

TEACHING NOTES

The key concepts can be posted in many different ways. If you don't want to use sentence sheets, here are some alternatives:

- Write the key concepts out on sentence strips.
- Write the key concepts out before class on a posted piece of butcher paper. Cover each concept with a strip of butcher paper and reveal each one as it is brought up in the class discussion.

SESSION 1.5 The Balloon-Rocket Mission

5. **Copy and prepare a set of the Student Balloon-Rocket Mission Graphing Sheets for each team.** Each team should have a set of the following sheets: Infrared, Ultraviolet, X-ray, and Gamma Ray. If you have decided to have students graph *all* of the electromagnetic energies, then also copy the Student Balloon-Rocket Mission Graphing Sheets for Radio, Microwave, Visible Light, and Solar Particles. (See the note on page 215.)

6. **Copy the Completed Balloon-Rocket Mission Graphing Sheets for classroom use.** Copy all eight Completed Balloon-Rocket Mission Graphing Sheets: Radio, Microwave, Visible Light, Infrared, Ultraviolet, X-ray, Gamma Ray, and Solar Particles. Have them ready to post on the Electromagnetic Spectrum chart as each one is discussed.

7. **Photocopy the Extension Graph of Solar Particles.**

GO! Introducing the Mission

1. **Another opportunity for students to be solar science experts.** Remind the class that last time they testified as solar-science experts for the prosecution in a trial against the Sun. Tell students that today they will be using their expertise again to advise four people about their exposure to various solar energies. Say that these four people are:

   - A student in a schoolyard.
   - A snowboarder on Mount Everest.
   - A high-altitude skydiver.
   - An International Space Station astronaut on a spacewalk.

2. **Each person is at a different altitude above sea level.** Tell the class that each person is located at a different altitude: the student is at 0 km, the snowboarder is at 7 km, the skydiver is at 40 km, and the astronaut is at 350 km. Ask students whether they think the amount of the Sun's energy that a person is exposed to changes with the person's altitude. Have students share and discuss their ideas with the class, or with a nearby classmate.

TEACHER CONSIDERATIONS

CD-ROM NOTES
The Balloon-Rocket Mission. Teachers can use this interactive to replace the transparency in the presentation of data for the balloon-rocket mission. To advance altitude and sequentially reveal the mission data to students, click the UP button above the altitude readings. To review the previous altitude, click the DOWN button. The data-bar graphs, and the digital readouts above them, will change based on the new altitude chosen. A specific altitude can also be selected by clicking on the desired altitude directly. The present altitude is shown by the position of the balloon-rocket and the digital readout at the top of the screen. Data graphs for radio, microwave, and solar particles are not shown because these values (shown at the bottom of the screen) remain the same for all altitudes. To enlarge the interactive to full screen, press CONTROL F for Windows and APPLE F for Macs. Click ESC to exit this mode. You can close the program any time, just as you would close a window on your desktop.

TEACHING NOTES
Snowboarding on Mount Everest? In 2001, the French snowboarder Marco Siffredi became the first person to climb Mount Everest and descend to base camp on a snowboard. Starting at the summit (8,848 meters), it took him two and a half hours to snowboard to base camp almost 2,448 meters below. (We know the brand of snowboard he rode, but the brand of sunscreen he wore has not been widely publicized.) Sadly, he disappeared while attempting a more difficult route down Mount Everest about a year later.

SESSION 1.5 The Balloon-Rocket Mission

3. **Reviewing which electromagnetic energies are dangerous to humans.** Refer to the posted chart of the Electromagnetic Spectrum from Session 1.3. Ask students which energies from the spectrum can be dangerous to humans. [Microwave, ultraviolet (or UV), X-ray, and gamma-ray radiation can all be harmful.]

4. **Launching a mission to measure energy amounts at different altitudes.** Tell students that today they will be conducting a mission simulation to measure solar-energy amounts at various altitudes. Using mission data, the class will then advise the four people about their exposure risk to dangerous energies.

5. **How data will be obtained.** Tell the class that in this imaginary mission, a balloon with a device attached to it will be used to measure the different kinds of energy from the Sun. Say that the device will radio back data to the class every 10 km as it rises higher and higher in the balloon above Earth.

6. **Using a rocket to gather data above Earth's atmosphere.** Ask the class if the balloon will be able to continue rising indefinitely, higher and higher into Earth's atmosphere. [No. A balloon is lifted by the atmosphere which surrounds it. At very high altitudes where the atmosphere is too thin, the balloon will not be able to rise any farther.] Explain that in this mission, when the balloon reaches 50 km, it will launch an attached rocket to carry the measuring device higher up. The device will radio back data every 50 km after the rocket is launched from the balloon.

7. **A mission like this has never happened in real life.** Explain that in real life, there has never been one single mission that has measured the Sun's energies at all altitudes. Instead, many different missions with balloons and rockets have made these measurements. The balloon-rocket simulation that the class is about to conduct is an imaginary all-purpose combination of several of these missions.

TEACHER CONSIDERATIONS

SESSION 1.5 The Balloon-Rocket Mission

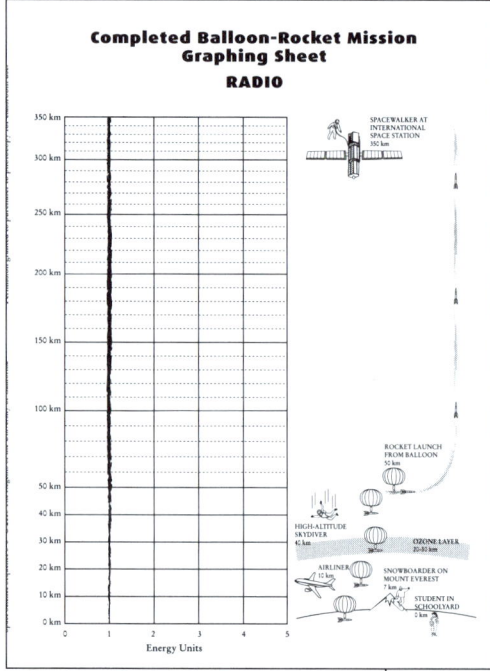

Preparing to Receive and Interpret Data

1. **Organize students into teams of four.** Tell the class that they will be working in teams of four. Explain that each team member will be responsible for graphing the data for one electromagnetic energy. Team members should help one another to record the data accurately, and they should also discuss their graphs with one another as the mission proceeds.

2. **Learning how to interpret the data first.** Tell teams that before launching the mission, you would like to show them some data that the balloon-rocket obtained earlier. Explain that the class will be looking at some graphs in order to learn how to correctly interpret data from the balloon-rocket.

3. **Show one of the static, or unchanging, completed graphs.** Choose one of the following three transparencies: Completed Balloon-Rocket Mission Graphing Sheets: Radio, Microwave, or Visible Light.

4. **Familiarizing students with the data-graph sheets.** First, point out the illustrations to the right of the graph itself. Point out the student, snowboarder, skydiver, and astronaut. Ask students to imagine the Sun's energy coming toward Earth from above. Say that the various people and altitudes pictured on the right correspond to the altitude levels on the vertical axis of the graph. Say that the graph's horizontal axis is in energy units. A higher level of energy units indicates a higher amount, or intensity, of an energy, while a lower level of energy units indicates a lower amount, or intensity, of energy. Ask the following questions to check for student understanding:

- What is the highest altitude shown on the graph? [350 km.]
- What is the lowest altitude shown on the graph? [0 km, which is the altitude at sea level.]
- At what altitude is the student in the schoolyard? [0 km.]
- What is the altitude of the snowboarder on Mount Everest? [7 km.]
- How high up is the airliner? [A little above 10 km.]
- Where is the high-altitude skydiver? [40 km.]
- At what altitude will the rocket launch from the balloon? [50 km.]
- How high up is the International Space Station? [350 km.]

TEACHER CONSIDERATIONS

TEACHING NOTES

Don't rush through this section—make sure students understand how to properly read the graphs before moving on.

Some additional options for this graphing activity:

1. Have teams work together to record data individually for the **same** energy. Afterward, assign students into new groups, with each student as the "expert" on a particular energy. This is an especially good option if you feel your students could help one another with their graphing.

2. For older or more advanced students, go over only one static graph briefly and then divide all eight energies among a team of four students, so that each student is responsible for creating two graphs. If you use this option, four of the sheets will be less interesting to graph than the other four. Distribute the sheets as suggested in the table below:

	Less Interesting		More Interesting
Student 1:	RADIO	and	INFRARED
Student 2:	MICROWAVE	and	ULTRAVIOLET
Student 3:	VISIBLE LIGHT	and	X-RAYS
Student 4:	SOLAR PARTICLES	and	GAMMA RAYS

SESSION 1.5 The Balloon-Rocket Mission

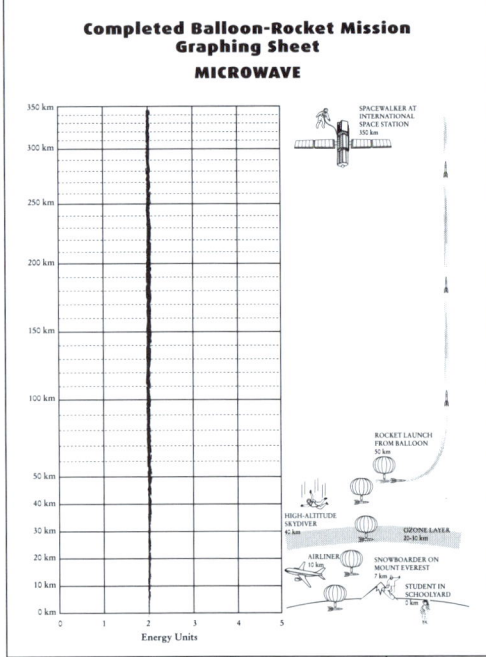

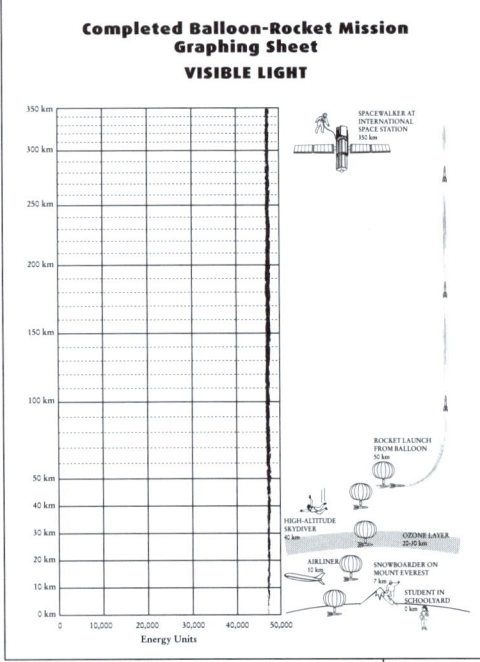

5. **How to interpret the graphed data.** Now have students look at the graphed data. Ask how many energy units there are at 0 km. Continue through different altitudes, asking students each time what the energy level is at that altitude. Once students begin to notice that the energy level remains the same for all altitudes on the graph, ask, "For this energy, is a person exposed to the same amount of it whether they are at sea level or on the International Space Station?" "Why?" [Yes. For this energy, a person is exposed to the same amount of energy at both altitudes, because the energy level does not change with altitude.] Make sure students understand that even though the energy level remains the same, that does *not* mean that the energy level is zero!

6. **Go through the remaining two static completed graphs.** Show the other two Completed Balloon-Rocket Mission Graphing Sheets transparencies and ask students questions about the remaining two graphs as you did with the first one. Make sure they understand *why* these data graphs indicate no change in energy.

7. **Energies that are blocked show no change.** Explain to the class that these energies—radio, microwave, and visible light—show no change because they are not being blocked by anything. These energies are the same at 0 km as at 350 km because nothing is preventing the energy from reaching Earth's surface. Share the following with the class:

   - **Radio:** Very little radio energy comes from the Sun, and there is nothing within 350 km of Earth that seems to block it.
   - **Microwave:** Like radio waves, there is no change in microwave levels with altitude.
   - **Visible light:** On a clear day, visible light from the Sun shines all the way to the surface of Earth without being blocked by anything.

8. **Post the copies of the three static completed graphs onto the Electromagnetic Spectrum chart from Session 1.3.** Add the Completed Balloon-Rocket Mission Graphing Sheets to the radio, microwave, and visible light regions of the spectrum. Say that later the class will be adding more graphs to the chart after they have finished collecting and analyzing data from the mission.

TEACHER CONSIDERATIONS

SESSION 1.5 The Balloon-Rocket Mission

9. **Pass out the Student Balloon-Rocket Mission Graphing Sheets.** Each student should receive a graphing sheet. As you pass out the graphing sheets, point out which energy (indicated on the sheets) each student will be graphing (infrared, ultraviolet, X-ray, or gamma ray). Explain that once everyone has completed their graphs, the class will be able to tell what kinds of solar energies make it to Earth's surface and how altitude affects the intensity of different energies. Remind teams that they should help each other with graphing, and that they should also discuss their graphs with one another.

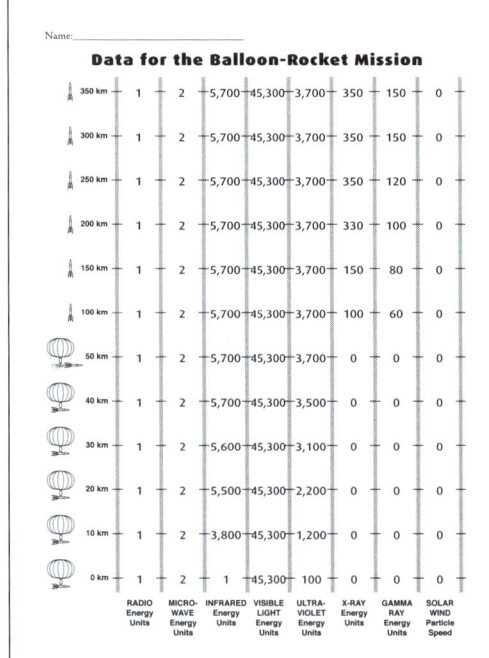

The Balloon-Rocket Mission

1. **Show the Data for the Balloon-Rocket Mission transparency, with a sheet of paper covering it.** Tell the class that they are now ready for the mission to begin! Say that the balloon-rocket is on the launch pad at 0 km, ready for launch. Explain that it is already transmitting data, and that the first data point students will be making will be for ground level, or 0 km. Move up the sheet of paper covering the data so that only the bottom row of data (for 0 km) is visible to the class.

2. **The data readout shows the amount of electromagnetic energy the balloon-rocket's device is measuring.** Explain to the class that the numbers shown indicate the amount, or intensity, of energy the device is currently measuring at 0 km. If students ask about the solar-wind data shown, say that they will be discussing that data a little later on.

3. **Help students to understand the data.** Ask students the following questions to help them build understanding: "Which energies of the Sun's output are blocked and not getting down to Earth's surface?" [X-ray and gamma ray.] "Which energies reach Earth's surface at very low levels or intensities?" [Radio, microwave, infrared, and UV.] "Which of these energies would be damaging to life if they reached 0 km at high intensity?" [Microwave, UV, X-ray, and gamma ray.] "Which energy is not blocked much at all at 0 km?" [Visible light.]

TEACHER CONSIDERATIONS

TEACHING NOTES

The CD-ROM can be used to show the balloon-rocket mission data.

Depending on your students' ages and the experience they have with graphing, you may need to spend some time helping them with recording the first few data points, until they get the hang of it.

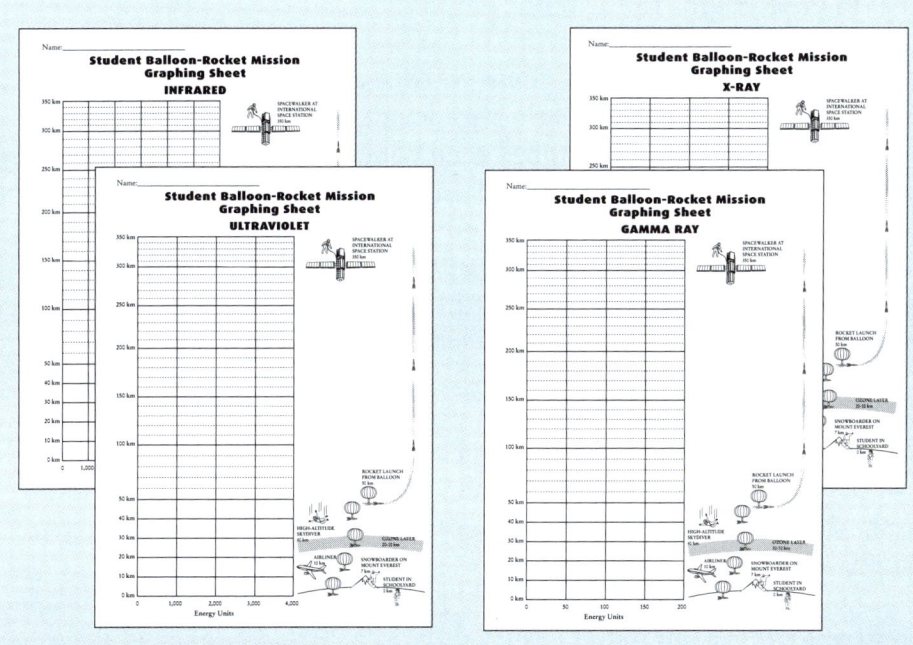

SESSION 1.5 The Balloon-Rocket Mission

4. **Show students how to record the data on their Student Balloon-Rocket Mission Graphing Sheets.** Remind students that the horizontal axis is in energy units, while the vertical axis is in altitude. Have students graph their data for 0 km. If your students need help with graphing, first have them find 0 km on the vertical axis, and then have them move along the horizontal axis to find the correct energy level for the energy that they are graphing. Say that they should make a dot at the spot that indicates the correct energy level *and* altitude. Teammates should check each other's data to make sure it was recorded accurately. Explain that later on, students will connect the dots on their graphs to form line graphs.

5. **Go over the affect on the student in the schoolyard.** Ask if any of the four people mentioned earlier are located at 0 km. [The student.] Ask if, based on the data that has been collected, the student in the schoolyard is being affected by any of the energies. [Yes. Visible light and some UV. The student is also exposed to some very small amounts of radio, microwave, and infrared.] Say that while visible light is not harmful, UV *can* be harmful, and in future sessions, the class will learn more about the dangers of exposure to UV.

6. **Reveal the data for 10 km.** Have the class count down to "launch!" and then move up the sheet of paper to reveal the data for 10 km. Call out "10 km" as you do so. Make sure students record their data on the 10 km-altitude line of their graphing sheets. Again, have team members check each other's work.

7. **Compare the 10-km data to the 0-km data.** Ask students which energies changed intensity between 0 and 10 km. [Infrared and UV.] Ask which energies remained the same at the two altitudes. [Radio, microwave, visible light, X-ray, and gamma ray.] Ask which energies the snowboarder is exposed to at 7 km. [Infrared, visible light, and UV. There are also small amounts of radio and microwave at 7 km.] Remind the class that UV can be harmful. Ask who is being exposed to more UV—the student or the snowboarder. [The snowboarder.]

8. **Continue calling out altitudes as you reveal each successive row of data.** Ask students to let you know when the balloon-rocket has reached the altitude of the skydiver. At 40 km, ask which energies have changed in intensity from 10 to 40 km. [Infrared and UV.] Ask if the skydiver should be concerned about his exposure to any of the solar energies as he begins his dive. [Yes. At 40 km, the skydiver is being exposed to much more UV than the snowboarder or student.]

One teacher said, "Some students told me, 'hot air rises' and wondered why the tops of mountains have snow. It was a great teachable moment. I loved the graphing and the discussion with this one."

TEACHER CONSIDERATIONS

TEACHING NOTES
In the next session, students will learn about the different forms of ultraviolet light—UV-A, UV-B, and UB-C.

SESSION 1.5 The Balloon-Rocket Mission

9. **At 50 km, announce that the balloon has gone as high as it can, and the rocket will carry the measuring device higher.** (You can have the students call out "rocket launch!" at 50 km.) Say that from now on, the device on the rocket will transmit data every 50 km, instead of every 10 km. Continue to reveal the rows of data as before, until you reach 350 km.

10. **The energy exposure at the altitude of the International Space Station.** At 350 km, ask which solar energies have changed in intensity from 40 to 350 km. [UV, X-ray, and gamma ray.] Ask if any of these energies could be harmful. [All of them.] Tell the class that in a future session, they will be reading about how International Space Station astronauts protect themselves from UV, X-ray, and gamma-ray exposure.

Interpreting the Data and Identifying Shields

1. **Have students connect their data points.** When students have finished recording the data for 350 km, have them connect their data points from 0 km up to 350 km to form line graphs.

2. **Optional: Discussing which energies changed.** If you had your students record data for *all eight graphs*, have them discuss in their teams which energies changed as the balloon-rocket mission gained altitude. Ask, "Which graphs changed with altitudes?" [Infrared, UV, X-ray, and gamma ray.]

3. **Discussing why energy levels change with altitude.** Have teams look over their completed infrared, UV, X-ray, and gamma-ray graphs. Ask the class whether these energies increase or decrease in intensity as altitude increases. [They all increase as altitude increases.] Ask next, "What might cause this to happen?" "Why might the intensity of an energy decrease below a certain altitude?" [At a certain altitude, something might be blocking the energy.]

4. **Introducing shields.** Tell the class that a *shield* is anything that blocks energy from the Sun. Say that the data they have collected from the balloon-rocket mission is evidence that there are shields at certain altitudes that block some solar energies. Post on the concept wall, under Key Space Science Concepts:

Shields at various altitudes prevent some solar energies from reaching Earth's surface.

TEACHER CONSIDERATIONS

TEACHING NOTES

Evidence and explanations. In Session 1.4, students learned about scientific evidence by using their research notes to support the finding that the Sun was responsible for the many mysterious events introduced at the beginning of the unit. In this session, students again gather evidence—this time during a mock mission. Students behave as scientists as they discuss their evidence and try to come up with explanations that best match all the evidence they have. They learn that a scientist's observations and findings must be critically analyzed, examined, and verified by other scientists before being considered valid scientific evidence.

Astronomers study extremely remote objects, so it's not common for them to have a physical specimen to show as evidence. Instead, the evidence they gather is frequently comprised of careful observations, images, measurements, and other data that can be verified by others.

SESSION 1.5 The Balloon-Rocket Mission

5. **Have students describe their findings for infrared, UV, X-ray, and gamma ray using their graphs.** Ask students to report on each type of energy and at what altitude its intensity begins to change. For each energy, also ask students what the implications are for the student, snowboarder, skydiver, and astronaut. Using the information below, explain what shields us from these solar energies:

- **Infrared:** Infrared energy levels are highest at and above 40 km. Below 40 km, Earth's atmosphere (mostly the water vapor in the atmosphere) absorbs infrared energy. The student in the schoolyard at 0 km is shielded from most solar infrared, while the snowboarder is shielded less. The skydiver at 40 km and the astronaut at 350 km, however, are not shielded by the atmosphere from infrared. However, infrared energy is not harmful.

- **Ultraviolet (or UV):** The ozone layer of Earth's atmosphere protects life on Earth from UV energy. The intensity of UV is greatly reduced as it comes toward Earth. Most of the reduction happens between 20 and 30 km, in the ozone layer. Ozone is a form of oxygen that absorbs UV energy. Ask students which of the four people (student, snowboarder, skydiver, or astronaut) is the most protected and which is the least protected from UV energy. [The student and the snowboarder are the most protected, while the skydiver and the astronaut are the least protected.]

- **X-ray and gamma ray:** These energies are almost completely blocked in the highest layers of the atmosphere. Ask if any of the four people should be concerned about X-ray or gamma-ray exposure. [Only the astronaut needs to worry about exposure to X-rays and gamma rays.]

6. **Post the four changing completed graphs onto the Electromagnetic Spectrum chart from Session 1.3.** Attach the Completed Balloon-Rocket Mission Graphing Sheets for infrared, UV, X-ray, and gamma ray to their regions of the Electromagnetic Spectrum chart.

Solar Particles and Earth's Magnetosphere

1. **One more graph to consider.** Tell the class that so far, the balloon-rocket mission has shown them how electromagnetic energies coming from the Sun vary according to altitude. Ask if electromagnetic energies are the only things that come toward us from the Sun. [No. The Sun also puts out solar particles.] If necessary, refer to the Particles from the Sun display sheet on the posted Electromagnetic Spectrum chart.

TEACHER CONSIDERATIONS

SESSION 1.5 The Balloon-Rocket Mission

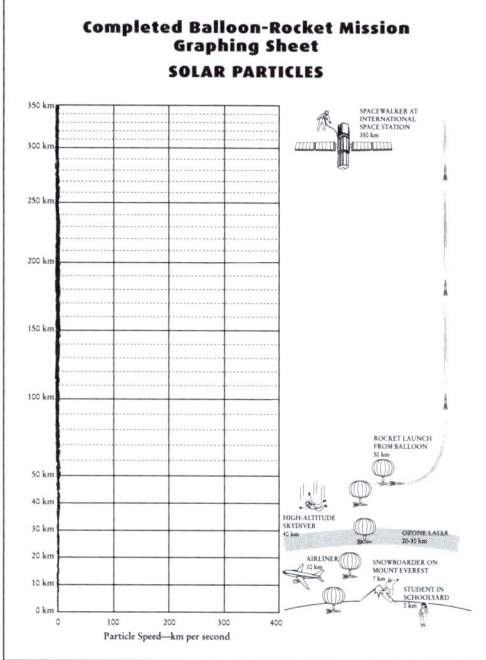

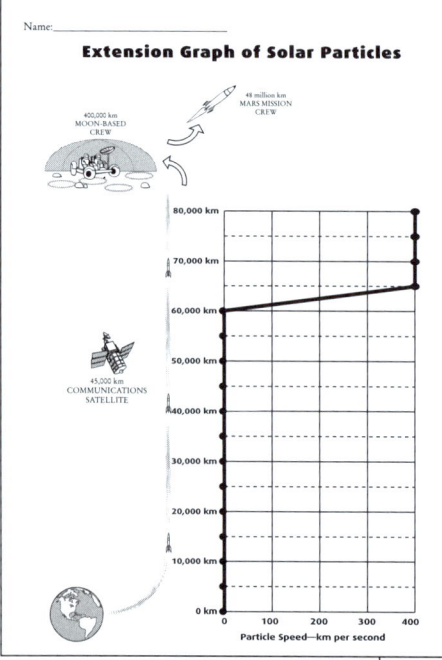

2. **Show the Completed Balloon-Rocket Mission Graphing Sheet: Solar Particles transparency.** Say that the balloon-rocket mission also gathered data for solar particles. Point out that for this graph, the horizontal axis is in particle speed instead of energy units. Explain that the idea is the same—this graph indicates the amount of solar particles at different altitudes. Ask students what the graph tells them. [The amount of solar particles does not change with altitude.]

3. **Show the Extension Graph of Solar Particles transparency.** Tell the class that the balloon-rocket mission only gathered data for solar particles up to 350 km. Say that this second graph shows measurements of solar particles from 0 km up to 80,000 km. Ask students if the graph indicates that there is a shield for solar particles. [Yes. A shielding effect occurs between 60,000 and 65,000 km.]

4. **Post copies of the solar-particles graph and its extension graph onto the Electromagnetic Spectrum chart from Session 1.3.** Attach the Completed Balloon-Rocket Mission Graphing Sheet: Solar Particles and the Extension Graph of Solar Particles onto the Electromagnetic Spectrum chart.

5. **Introduce the magnetosphere.** Explain that we are shielded from solar particles by Earth's magnetic field. The pattern of magnetism around Earth is called the *magnetosphere* (mag-NEET-oh-sfeer). Say that the magnetosphere steers solar particles away from Earth's surface, except near the North and South poles. Post on the concept wall, under Key Space Science Concepts:

   The magnetosphere shields Earth from solar particles at a very high altitude.

6. **Remind students that during a solar storm, solar flares and coronal mass ejections release gusts of solar particles.** Refer to the posted key concepts from Session 1.4:

Solar flares and coronal mass ejections (CMEs) occur during solar storms, when the Sun is active.

A solar flare releases large amounts of electromagnetic energies and solar particles into space.

A CME ejects particles and material from the Sun's corona at high speeds into space.

TEACHER CONSIDERATIONS

TEACHING NOTES

As students will find out soon (in Session 1.8), the International Space Station is shielded from solar particles on a normal day when the Sun is quiet. However, during a solar storm, the International Space Station may be exposed, and astronauts must take extra precaution to protect themselves.

SESSION 1.5 The Balloon-Rocket Mission

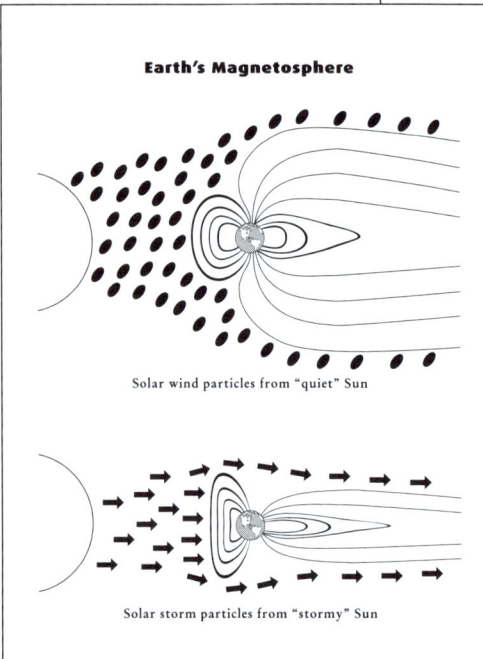

7. **Show the Earth's Magnetosphere transparency.** Have the class observe what happens to the Earth's magnetosphere when the Sun is active. Note that the magnetosphere gets distorted, allowing solar particles to get nearer to Earth. Explain that when this happens, satellites and International Space Station astronauts are temporarily outside the shield of the magnetosphere and can be affected by these energetic particles. Even those on Earth can be affected by energetic solar particles during a solar storm. Say that Earth's changing magnetic field can cause power surges in large electrical systems. Post on the concept wall, under Key Space Science Concepts:

The changing shape of Earth's magnetosphere during a solar storm can leave satellites and astronauts unshielded.

8. **Looking again at the calendar of mysterious events from Session 1.1.** Have the class look over the posted week-long calendar of mysterious events. Ask if any of the events listed could be linked to solar particles released during a solar storm. [Yes. The disabled TV, navigation, and cell-phone satellites, as well as the blackouts and unusual clouds of light and color could all be caused by solar particles.] If students have difficulty answering this question, ask them what they've learned about how long it takes particles from the Sun to reach Earth. [Since particles travel much more slowly than EM energies, it takes them two to three days after leaving the Sun to reach Earth. So the particles released from a large solar flare occurring on April 21 would not reach Earth until April 23 or 24.]

9. **Auroras are caused by solar particles.** Explain that the strange lights in the sky, called *auroras*, are indeed a result of particles from the Sun. When the Sun is quiet, they are seen faintly and only far to the north, but during a solar storm they are brighter and can be seen farther south, even in the southern United States!

10. **Drawing some conclusions.** Ask the class to think back to the student, snowboarder, skydiver, and astronaut. Discuss the exposure risks for each person from both EM energies and solar particles.

11. Save the Electromagnetic Spectrum chart with attached completed graphs for use again in the next session.

TEACHER CONSIDERATIONS

SCIENCE NOTES

Clarifying note: The balloon-rocket-mission data does not show any solar particles below 350 km. In reality, however, some of these charged particles do get funneled down toward the poles, hitting Earth's atmosphere, and causing auroras.

SESSION 1.6
Investigating Ultraviolet Shields

Overview

In the previous session, students learned that Earth's ozone layer does not entirely shield Earth from all ultraviolet energy, and that it is less effective at high altitudes. To determine which materials function effectively as ultraviolet shields, in this session students run experiments to test various materials. Conducting effective tests requires trial and error, so the ultraviolet-shielding investigation requires two sessions to complete. High student interest and students' gain in experimental-design experience make this activity worth the extra time. Also in this session, the class learns more about ultraviolet energy and its hazards. Using UV beads, which change color when exposed to ultraviolet light, student teams design and carry out experiments to test various shield materials. The importance of using a control and taking careful observation notes is emphasized as students conduct their tests. During this session, the key concepts that will be added to the classroom concept wall are:

- *Everyone, especially those at high altitudes, should be concerned about ultraviolet radiation from the Sun.*
- *Earth's ozone layer shields us from some of the Sun's ultraviolet energy.*

Investigating Ultraviolet Shields	Estimated Time
Ultraviolet Energy and Shields	15 minutes
Experimenting with UV Beads	10 minutes
Testing Various Shields	20 minutes
Total	**45 minutes**

Unit Goals

The Sun is a star, and a main source of energy for Earth.

The Sun gives off the full spectrum of electromagnetic energies, as well as solar particles.

The Sun's energy and matter output varies and is not constant.

Earth has protective shields located at various altitudes that help to block much of the Sun's harmful output from reaching Earth's surface.

Safety is a concern — without Earth's shields to protect us, some of the Sun's energies can be harmful.

What You Need

For the class:
- ❑ prepared key concept sheets from the copymaster packet or CD-ROM file
- ❑ the Electromagnetic Spectrum chart with all eight Completed Balloon-Rocket Mission Graphing Sheets attached (from Sessions 1.3 and 1.5)
- ❑ a 3-ft wide by 4-ft long sheet of butcher paper
- ❑ a marker
- ❑ assortment of possible shields for testing: different sunscreens, sunglasses, fabric swatches, lip balm, lotions, various papers, newspaper, waxed paper, aluminum foil, clear plastic cups with water
- ❑ (optional) four or more fluorescent UV lights or black lights
- ❑ (optional) extension cords
- ❑ (optional) 1 handheld electronic UV meter/sensor

TEACHER CONSIDERATIONS

TEACHING NOTES

If you have a limited amount of time, Sessions 1.6 and 1.7 can be condensed into one session. However, we strongly advise against this as students really enjoy running the UV-shielding experiments, and the activity was written with two sessions in mind.

The optional UV meter/sensor listed could be a nice add-on for Sessions 1.6 and 1.7. Several manufacturers make this type of handheld electronic device, which typically costs $20–25. (See the Resources and References section for ordering information.) These UV meters/sensors give instant digital measurements of UV-intensity levels, as quantified by the UV index (http://en.wikipedia.org/wiki/UV_index), an international standard linear-scale measurement used in daily weather forecasts and aimed at the general public. In the sessions here, the UV meter can be used wherever UV beads are used (i.e., behind potential shields or fully exposed to solar UV energy). The main advantage of an electronic UV monitor will be its quantitative measurements, which the beads cannot provide directly (although timing of their color changes is encouraged). You may want to use the UV meter/sensor *after* students have completed their own UV investigations in Session 1.7 and/or use it as a teacher resource to monitor solar UV intensity (e.g., compare clear and cloudy days; compare intensity over different times of day when the Sun is at different heights in the sky). Time permitting, or as a Providing More Experience activity, you could have students come up with new experimental designs using a UV meter in place of UV beads.

Key Vocabulary

Scientific Inquiry Vocabulary

Control
Evidence
Model
Observation
Prediction
Scale
Scale model
Scientific explanation

Space Science Vocabulary

Coronal mass ejection (CME)
Electromagnetic (EM) energy
Magnetosphere
Matter
Particle
Shield
Solar flare
Solar particle
Solar wind
Spectrum
Star
Ultraviolet (UV)

SESSION 1.6 Investigating Ultraviolet Shields

For each team of 2–3 students:
- ❏ 4–6 (or more) UV beads
- ❏ 1 dark, opaque film canister or other suitable container (to keep the beads shielded from UV)
- ❏ a few plastic sandwich bags (to cover bead when testing sunscreen or lotions)

For each student:
- ❏ 1 copy of the Ultraviolet-Shielding Experiments student sheet from the copymaster packet or CD-ROM file

Getting Ready
Before the day of the activity:

1. **Obtain UV beads.** (See the Resources and References section for ordering information.) These beads change color in ultraviolet light. If the beads are in assorted colors, expose them to sunlight and sort them out by color. (Students will be able to do controlled tests only if the beads they use in their teams are all the same color.) Place four to six beads into each film canister.

2. **Gather together a class set of objects and materials to test as shields.** For sunscreens, choose a variety of brands and SPF ratings.

3. **Determine where students will conduct their experiments.** We strongly recommend that you use sunlight as a source of ultraviolet energy. Outdoors is best, with a shady or indoor area for students to plan and prepare their shielding tests. A classroom with large windows that will let sunshine in during the session may be suitable. Even on an overcast day, there is enough ultraviolet light to do the activity.

4. **If the class will be staying indoors, decide where you will be setting up the black lights.** If you decide to use black lights, figure out where you will plug them in and if you will need extension cords. Choose locations around the classroom where students will have easy access to the lamps.

On the day of the activity:

1. **Prepare the key concept sheets.** Make a copy of each key concept and have them ready to post onto the classroom concept wall during the session.

2. **Decide how you will divide the class into teams of two to three.** If possible, have students work in pairs rather than in teams of three.

Use only fluorescent ultraviolet lamps. Incandescent ultraviolet lamps are extremely hot and are a risk for burns and fires.

TEACHER CONSIDERATIONS

TEACHING NOTES

The key concepts can be posted in many different ways. If you don't want to use sentence sheets, here are some alternatives:

- Write the key concepts out on sentence strips.
- Write the key concepts out before class on a posted piece of butcher paper. Cover each concept with a strip of butcher paper and reveal each one as it is brought up in the class discussion.

SESSION 1.6 Investigating Ultraviolet Shields

3. **Prepare the Ultraviolet Information Chart on a large sheet of butcher paper.** Copy the following chart:

Ultraviolet Information Chart			
Type of UV	UV-A	UV-B	UV-C
Uses	Skin tanning, black lights, "invisible" stamps.	Stimulates the skin to produce vitamin D.	Sterilization, disinfection.
Harmful Effects	May contribute to skin cancer and aging of cells.	Sunburn, skin cancer, eye damage, such as cataracts.	Lethal to all living things.
Shielding	Most blocked by stratosphere.	Most blocked by stratosphere.	All blocked by stratosphere.

4. **Post the Ultraviolet Information Chart and the Electromagnetic Spectrum chart from previous sessions.**

5. **Copy the Ultraviolet-Shielding Experiments student sheet.** Each student should have his or her own sheet. Alternatively, you could have students make their own data sheets.

6. **Set out the shield materials.** Place them in an area where students will have easy access to them during the session, or decide how you will carry everything outside to the testing area if students will be doing the experiments outdoors.

7. **If students will be testing indoors, set up the black lights around the classroom.**

TEACHER CONSIDERATIONS

SESSION 1.6 Investigating Ultraviolet Shields

GO! Ultraviolet Energy and Shields

1. **Learning more about UV energy today.** Remind the class that last time, they graphed and analyzed data collected from a balloon-rocket mission. The data showed them how the intensity of various EM energies changed according to altitude. Say that today, students will learn more about UV energy, its harmful effects, and what materials might be good shields to protect against UV exposure.

2. **Reviewing what a shield is.** Ask students what a shield is. [A shield is anything that blocks energy from the Sun.]

3. **Reviewing the UV-exposure data from the balloon-rocket mission.** Point to the Completed Balloon-Rocket Mission Graphing Sheets on the Electromagnetic Spectrum chart and ask the class, "Which energies would a student at sea level (or 0 km) be exposed to?" [Radio, microwave, infrared, visible light, and UV.] Ask, "Which of these energies might be dangerous to the student?" [UV is the only *energetic* electromagnetic energy that the student would be exposed to, and it can be dangerous.] Ask, "Who would be exposed to more UV light than the student?" [People at higher altitudes, such as the snowboarder, skydiver, and astronaut.] Post on the concept wall, under Key Space Science Concepts:

    Everyone, especially those at high altitudes, should be concerned about ultraviolet radiation from the Sun.

4. **Sharing about UV.** Ask students to share what they have heard or know about UV light, particularly any possible hazards associated with it. [UV can cause skin cancer, sunburn, eye damage (cataracts), etc.] Students might also mention some positive uses for UV, such as detecting counterfeit money, verifying ID cards, or special lighting effects.

TEACHER CONSIDERATIONS

SESSION 1.6   Investigating Ultraviolet Shields

5. **The importance of Earth's ozone layer.** Ask, "What shields us on Earth from most solar UV?" [The ozone layer shields us from most of the UV energy coming from the Sun.] Ask the class if they have heard of a problem with Earth's ozone layer. (Some students may have heard that it is getting thinner, or that there's a hole in it.) Say that some gases produced from modern technologies (such as air conditioners and aerosol cans) destroy ozone and have caused the ozone layer to become thinner over some parts of Earth. Explain that although this has been happening for only the past 50 years or so, it is a serious problem that many people are working on. Emphasize that the ozone layer in Earth's atmosphere serves an important function as a shield that protects us against harmful UV energy. Post on the concept wall, under Key Space Science Concepts:

   The Earth's ozone layer shields us from some of the Sun's ultraviolet energy.

6. **Going over the three types of UV energy.** Point to the prepared and posted Ultraviolet Information Chart and tell students that there are three different kinds of UV: A, B, and C. Say that the different energies in the UV part of the electromagnetic spectrum are just like the different colors in the visible part of the spectrum. Explain that just as red light has the least energetic photons and violet light the most energetic photons in the visible spectrum, UV-A has the least energetic photons and UV-C the most energetic photons in the ultraviolet part of the spectrum.

7. **Discussing the Ultraviolet Information Chart.** Go over the information on the chart briefly with students. Ask, "Which type of UV is most harmful to life?" [UV-C.] Say that we're fortunate that ozone is especially good at absorbing both UV-B and UV-C. Tell students that some UV-A and UV-B from the Sun does reach the ground, though. Ask, "What are the harmful effects of UV-A and UV-B?" [UV-A may contribute to skin cancer and cause cells to age. Too much UV-B causes sunburn and skin cancer and can contribute to eye damage, such as cataracts.] Emphasize that UV rays can even penetrate an overcast sky—we are exposed to UV even on cloudy days!

8. **Some exposure to UV is beneficial.** Ask, "Would it be best if we were never exposed to any UV?" [No. People need vitamin D from UV, and some sun exposure is healthy.] Conclude that people do need some exposure to UV, but too much unprotected exposure can be harmful.

One teacher said, "This session really got the kids thinking about labels on materials and also how important sun block and other blocking shields are to protect people from the harmful effects of the Sun. I had the kids take notes on the different types of UV rays and they put them in their journals. We talked about possible blockers and how bad the UV rays can be. They were surprised by the results on some of the materials."

TEACHER CONSIDERATIONS

SCIENCE NOTES
UV-B also destroys folate, an essential nutrient for reproduction.

SESSION 1.6 Investigating Ultraviolet Shields

9. **Discussing possible UV-shield materials.** Ask, "What can we do to protect ourselves from being exposed to too much UV?" [Use sunscreen; wear sunglasses, hats, and layers of clothing; stay out of the midday sun, etc.]

Experimenting with UV Beads

1. **Experiments to test possible shields.** Tell the class that they will be designing and conducting some experiments to determine what materials work best to shield against UV. Call on students to list some ideas about the materials they might test.

2. **Using a detector.** Tell students that since UV light cannot be seen, they will need some kind of detector that can indicate whether or not there is UV light present.

3. **Hold up a UV bead.** (Be sure to keep it out of the sunlight!) Ask students to describe its color. Say that the bead will change color if exposed to ultraviolet light.

4. **Demonstrate how a UV bead works.** Ask students to observe carefully as you expose the bead to some UV light. Move near a window in the classroom or use a black light to demonstrate the color change. Be prepared for some oohs and aahs! Say that the bead will become colorless again (or recover from its "sunburn") if it is kept out of UV light for a few minutes.

5. **How students might test a shield.** Hold up a possible shield (such as a piece of fabric or a pair of sunglasses) and ask students for ideas about how they would use the bead and shield to test the shield's effectiveness.

6. **Emphasize the importance of using a *control* bead.** If students have not yet brought up the idea of using a *control* bead, be sure to bring it up now. Ask the class how they can be sure that the bead's color change (or lack of it) is really due to the shield itself. Explain that that is why a control is necessary in scientific experiments. Say that in order to test the effect of a shield on a bead, students must see what happens to a bead that *does not have a shield*. Say that for each experiment they conduct, students must use two beads—a test bead (exposed to UV while protected by a shield) and a control bead (exposed to UV without the protection of the shield). This way, students can evaluate the shield's effectiveness accurately by comparing the color change and the speed of the color change between the test bead and the control bead.

TEACHER CONSIDERATIONS

TEACHING NOTES

If you have beads that turn red, you might say that the beads will become "sunburned" when exposed to too much UV. This can help students to visualize what is going on, but be sure to mention that what causes the bead to become red is nothing like what happens to skin when it becomes sunburned.

The use of a control bead is extremely important—be sure to clearly emphasize to your students that they must use a control bead with each of their experiments. (Note: If you have a set of assorted colors of UV beads, be sure to sort the beads by color first. Students should use beads of the same color when conducting their experiments.)

SESSION 1.6 Investigating Ultraviolet Shields

7. **Beads can change color very quickly.** Warn students that in strong sunlight, the beads can change color very quickly. Say that students should observe any changes very carefully. By seeing whether a bead changes color, and how quickly it does so (as compared to the control bead), students will be able to evaluate the effectiveness of various test shields.

Testing Various Shields

1. **Divide students into teams of two to three.**

2. **Pass out the Ultraviolet-Shielding Experiments student sheets, or have students make their own data sheets.** Go over the chart with the students. Say that for each experiment they conduct, they should take careful notes about what shield they tested, how they tested it, and what they saw.

3. **Show the class all of the materials they can use for their experiments.** If watches or stopwatches are available, tell students they may use them for their tests. Say that to test sunscreen, students should place the bead in a plastic sandwich bag and apply the sunscreen to the bag, *not* the bead. Point out the film canisters and tell students they should use them to protect their beads from UV until they are ready to begin their experiments.

4. **Students may have to test a shield more than once.** Let students know that they may have to try more than once to come up with an effective test for each shield. On their data sheets, they should record how effective a test was, and any changes they would do next time to more effectively test the shield.

5. **Have student teams plan what materials they would like to test.** Give teams a few minutes to discuss which materials they would like to test and how they plan to test each shield.

6. **Review important points with the class.** Say that students should work carefully. Let them know they will have part of another class session to continue testing shields, so they should use today to learn how to conduct their tests properly. Emphasize again the importance of taking good observation notes and of using a control bead in their experiments.

7. **Move the class outside to the testing area.** Point out areas of plentiful sunshine as well as shaded areas where students can set up their experiments. If the class will be staying indoors, point out where you have set up the black lights.

TEACHER CONSIDERATIONS

SESSION 1.6 Investigating Ultraviolet Shields

8. **Pass out the beads in the film canisters.** Caution students to keep their beads in the film canisters until they are ready to expose them to UV light. Remind them that any exposed beads will have to be kept in the dark for a few minutes before they can be used again.

9. **Allow students to begin their experiments.** Circulate among teams, reminding students to use control beads and to take careful notes. Ask teams how they are testing different shields and offer advice if needed.

10. **Ten minutes before the session ends have students begin to clean up.** Have them save their Ultraviolet-Shielding Experiments student sheets or collect them to redistribute in the next session. Return to the classroom.

11. **More testing next time.** Ask teams to briefly share some of the results of their testing—what worked and what didn't work. Say that the class will continue testing shields next time. Ask students if there are any materials they did not have today that they might want to test next time. [Makeup, a favorite brand of sunscreen or suntan oil, etc.] Have students predict whether or not they think these materials would be good UV shields. Invite students to bring in materials from home to test during the next session.

TEACHER CONSIDERATIONS

SESSION 1.7

Concluding the Ultraviolet-Shields Investigation

Overview

In the previous session, student teams began their ultraviolet-shielding experiments. It is likely that most teams learned more about the obstacles of doing an effective test than which materials are effective shields. In this session, teams begin by discussing their experiment successes and mistakes with one another. Students are encouraged to learn from one another's mistakes and are given additional time to continue their shielding experiments. The session ends with a class discussion of experiment results, and students are asked to assess the ultraviolet-exposure risk of four people located at different altitudes. During this session, the key concept that will be added to the classroom concept wall is:

- *Various materials can shield a person from UV radiation, but some shields are more effective than others.*

Concluding the Ultraviolet-Shields Investigation	Estimated Time
Sharing Experiment Methods	5 minutes
Continuing the UV-Shield Experiments	20 minutes
Discussing Experiment Results	15 minutes
Advising on UV Protection	5 minutes
Total	**45 minutes**

What You Need

For the class:
- ❏ overhead projector or computer with large-screen monitor or LCD projector
- ❏ prepared key concept sheet from the copymaster packet or CD-ROM file
- ❏ any materials students have brought from home to test
- ❏ all materials from Session 1.6
- ❏ transparencies of the Pre-unit 1 Questionnaire from Session 1.2 (three pages)

For each student:
- ❏ the Ultraviolet-Shielding Experiments student sheet from Session 1.6

Getting Ready

Prepare the key concept sheet. Make a copy of the key concept and have it ready to post onto the classroom concept wall during the session.

Unit Goals

The Sun is a star, and a main source of energy for Earth.

The Sun gives off the full spectrum of electromagnetic energies, as well as solar particles.

The Sun's energy and matter output varies and is not constant.

Earth has protective shields located at various altitudes that help to block much of the Sun's harmful output from reaching Earth's surface.

Safety is a concern — without Earth's shields to protect us, some of the Sun's energies can be harmful.

TEACHER CONSIDERATIONS

TEACHING NOTES

The key concepts can be posted in many different ways. If you don't want to use sentence sheets, here are some alternatives:

- Write the key concepts out on sentence strips.
- Write the key concepts out before class on a posted piece of butcher paper. Cover each concept with a strip of butcher paper and reveal each one as it is brought up in the class discussion.

Key Vocabulary

Scientific Inquiry Vocabulary

Control
Evidence
Model
Observation
Prediction
Scale
Scale model
Scientific explanation

Space Science Vocabulary

Coronal mass ejection (CME)
Electromagnetic (EM) energy
Magnetosphere
Matter
Particle
Shield
Solar flare
Solar particle
Solar wind
Spectrum
Star
Ultraviolet (UV)

SESSION 1.7 Concluding the Ultraviolet-Shields Investigation

GO! Sharing Experiment Methods

1. **Have students sit in their teams from the previous session.** Pass out the Ultraviolet-Shielding Experiments student sheets if you collected them, or have students take out these sheets.

2. **Reviewing what worked and didn't work last time.** Ask teams to share what difficulties they had in conducting their tests with particular shield materials. Also ask teams to share what procedures were effective for testing shields. Some common difficulties that students may bring up:

   • Exposing the beads to sunlight unintentionally.
   • Exposing beads for so long in bright light that it is difficult to distinguish the change of the test bead from the control bead. (Note: This could also just mean that the shield is not effective.)
   • Having difficulty controlling the length of exposure of the beads accurately.
   • Observing changes in the test beads under very bright conditions when the beads change color very quickly.

3. **Suggestions for improving experiments.** Have teams share suggestions on how to deal with some of the difficulties that were discussed. Tell students they should use one another's tips to help them improve how they conduct their UV-shield experiments.

4. **Sharing shield materials brought from home.** Invite students who brought materials to test to show them to the class. If time allows, have students predict which materials they think would be good (or poor) shields.

Continuing the UV-Shield Experiments

1. **Review important experiment points with the class.** Emphasize that students should take careful observation notes. Also remind them of the importance of using control beads.

2. **Return to the testing area.** Bring all shield materials and have teams continue with their tests.

3. **Pass out the beads in the film canisters.** Circulate among teams, helping or offering advice as needed. Remind teams to record their experiments on their data sheets.

4. **Twenty-five minutes before the end of the session, have students begin to clean up.** Return to the classroom.

TEACHER CONSIDERATIONS

SESSION 1.7 Concluding the Ultraviolet-Shields Investigation

Discussing Experiment Results

1. Drawing conclusions about the shield materials tested. Have the class discuss which shields they found to be effective or not effective in blocking UV energy. Some possible questions to bring up with students:

- Did their experiments answer any specific questions they had about what makes something a good shield for UV light?

- Were they surprised by any of their results? Were there shields they thought would work but did not? Were there shields that worked better than students thought they would?

- Were there any materials that worked only under specific conditions? (For instance, did the effectiveness of fabric change if it was wet?)

- Which materials were the best UV shields? Which were the worst?

After the discussion, post on the concept wall under Key Space Science Concepts:

Various materials can shield a person from UV radiation, but some shields are more effective than others.

2. Optional: Have teams that tested materials from home share their findings. If time allows, have teams that tested materials from home share their results with the class. Survey the class to see if their predictions were correct or not!

Advising on UV Protection

1. Looking again at the Completed Balloon-Rocket Mission Graphing Sheet for ultraviolet light. Ask students to compare the UV-exposure risks for the student, snowboarder, skydiver, and astronaut. [The astronaut is exposed to much more UV than the student. UV exposure increases with altitude.]

2. Students use their expertise about UV energy and shields to advise the four people. Tell the class that they will be using their newly gained expertise to write a letter to the four people, advising them about the risks of UV exposure and what shields they should use to protect themselves.

One teacher said, "There was a period when some clouds passed over, and we talked about the presence of UV rays even then. I think the students learned a lot from this lesson about UV rays and the need to protect themselves."

TEACHER CONSIDERATIONS

TEACHING NOTES
If time is short, one option would be to give the Advising on UV Protection activity as a homework assignment.

SESSION 1.7 Concluding the Ultraviolet-Shields Investigation

3. **Let students know what to include in their letters.** Ask students to provide the following information in their letters using evidence from the balloon-rocket mission, their UV-shielding experiments, as well as any class charts:

   • How much UV is each person exposed to?
   • How can UV be harmful to them?
   • What can they do to protect themselves from too much UV?
   • **Extra credit**: Explain each person's risk from X-ray, gamma-ray, or solar-particles exposure. During a solar storm, might the person be exposed to more X-rays, gamma rays, or solar particles?

4. **Review Question #s 2, 3, and 5 from the Pre-unit 1 Questionnaire with students.** Show the transparencies of the questionnaire. Have students look again at Question #s 2, 3, and 5 and ask how they would answer these questions now.

TEACHER CONSIDERATIONS

ASSESSMENT OPPORTUNITY
EMBEDDED ASSESSMENT: UV PROTECTION LETTERS
If you choose to use the letter-writing activity as an embedded assessment, please see the scoring guide on page 89 of the Assessment section.

QUESTIONNAIRE CONNECTION
Use this opportunity to review Question #s 2, 3, and 5 on the questionnaire with your students. If necessary, review with students the difference between particles and electromagnetic energies.

SESSION 1.8
Living with a Stormy Sun

Overview

The Post-unit 1 Questionnaire, which students take in this session, has the same questions as in the pre-unit questionnaire, although arranged in a different order. The conclusion of this unit with the post-questionnaire allows students to gauge how much they have learned about solar science. In this session, students review what they have learned about the possible terrestrial effects of a solar storm. A student reading about astronauts on the International Space Station teaches students how astronauts protect themselves from exposure to dangerous energies from the Sun. With the case of the mysterious events finally solved, students finish the unit by taking the post-unit questionnaire. During this session, the key concept that will be added to the classroom concept wall is:

- *We must all be concerned about bursts of energy from the Sun.*

Living with a Stormy Sun	Estimated Time
The Importance of Solar Science	10 minutes
Space Weather and the International Space Station—Student Reading	10 minutes
Taking the Post-unit 1 Questionnaire	25 minutes
Total	**45 minutes**

What You Need

For the class:
- ❑ prepared key concept sheet from the copymaster packet or CD-ROM file

For each student:
- ❑ 1 copy of the Space Weather and the International Space Station—Student Reading (two pages) from the copymaster packet or CD-ROM file
- ❑ 1 copy of the Post-unit 1 Questionnaire (four pages) from the copymaster packet or CD-ROM file

Getting Ready

1. **Prepare the key concept sheet.** Make a copy of the key concept and have it ready to post onto the classroom concept wall during the session.

2. **Make copies of student reading and questionnaire.** For each student, make a copy of the Space Weather and the International Space Station—Student Reading (two pages) and a copy of the Post-unit 1 Questionnaire (four pages).

Unit Goals

The Sun is a star, and a main source of energy for Earth.

The Sun gives off the full spectrum of electromagnetic energies, as well as solar particles.

The Sun's energy and matter output varies and is not constant.

Earth has protective shields located at various altitudes that help to block much of the Sun's harmful output from reaching Earth's surface.

Safety is a concern — without Earth's shields to protect us, some of the Sun's energies can be harmful.

TEACHER CONSIDERATIONS

TEACHING NOTES

The key concepts can be posted in many different ways. If you don't want to use sentence sheets, here are some alternatives:

- Write the key concepts out on sentence strips.
- Write the key concepts out before class on a posted piece of butcher paper. Cover each concept with a strip of butcher paper and reveal each one as it is brought up in the class discussion.

By the end of this unit, your classroom wall should be covered by many pieces of information posted during the sessions: the key concepts on the concept wall, the Electromagnetic Spectrum chart, information about ultraviolet light, and balloon-rocket mission graphs that give evidence for Earth's protective shields. You can choose whether or not to leave these materials posted during this final session when students take the Post-unit 1 Questionnaire. Here, however, are some good reasons to leave these pieces of information up:

None of these posted materials is a direct answer to the questions on the questionnaire. To answer the questions effectively, students must select which concepts are relevant and apply them correctly. Doing so demonstrates understanding, regardless of whether or not some related concepts or information are posted. Selecting and applying information builds learning and reinforces concepts during this assessment process, giving you a more knowledgeable group of students in the long run.

Key Vocabulary

Scientific Inquiry Vocabulary

Control
Evidence
Model
Observation
Prediction
Scale
Scale model
Scientific explanation

Space Science Vocabulary

Coronal mass ejection (CME)
Electromagnetic (EM) energy
Magnetosphere
Matter
Particle
Shield
Solar flare
Solar particle
Solar wind
Spectrum
Star
Ultraviolet (UV)

SESSION 1.8 Living with a Stormy Sun

GO! The Importance of Solar Science

1. **Review what students have learned about the Sun's energy and its effects on Earth.** Ask students to consider everything they have learned about the Sun in this unit. Ask, "Why is solar science important to everyone on the planet?" [We need to learn more about the Sun's energy output and when it can rise to dangerous levels for life on Earth. We need to learn about shields such as the magnetosphere, the ozone layer in the atmosphere, sunscreens, and other protective shields.]

2. **Discussing the harmful effects of the Sun.** Ask students what harmful effects can come from the Sun's output. [Satellites can be damaged, and electrical power systems can be disrupted by solar storms. Space travelers are exposed to high levels of dangerous radiation not experienced by those on Earth's surface. Changes in the atmosphere—such as the thinning of the ozone layer—allow more harmful ultraviolet energy to reach Earth's surface.]

3. **The conclusion of the mystery.** Congratulate the students on their work as scientific "detectives," and say that they have successfully identified the Sun as the cause of the mysterious events brought up in Session 1.1.

Space Weather and the International Space Station—Student Reading

1. **Pass out the Space Weather and the International Space Station—Student Reading to each student.** Give students about 10 minutes to read it.

2. **Discuss the reading.** Ask students what questions they have about the reading. Discuss how astronauts react and protect themselves from an active or stormy Sun.

3. **After the discussion, post on the concept wall under Key Space Science Concepts:**

   We must all be concerned about bursts of energy from the Sun.

TEACHER CONSIDERATIONS

Name:_____

Space Weather and the International Space Station—Student Reading

We all know that life on Earth could not exist without the Sun's energy. But some of the electromagnetic energy and particles coming from the Sun can be harmful to living things. Fortunately for us here on Earth, we are shielded from most of the Sun's harmful energies. Earth's atmosphere protects us from solar gamma rays, X-rays, and most solar ultraviolet light. Earth's magnetosphere shields us from solar energetic particles.

In the past, people weren't too concerned about harmful solar energies because of our planet's shields. Even when huge solar storms sent extra bursts of energy our way, this was just "space weather" and didn't affect us too much. But now, with space travel and satellites and other new technology, we all need to be concerned about space weather.

The crew of the International Space Station (ISS) live and work in space for months at a time. They orbit at 340 to 420 km above Earth. They fly above most of the atmosphere, so they need special protection from UV, gamma rays, X-rays, and solar particles.

Scientists and engineers work hard to make sure the crew of the ISS will be safe. They built the walls of the ISS with extra shielding. They plan how high the station will orbit, based on the Sun's cycle. About every 11 years, solar storms are more frequent. This is called *solar maximum*. During solar maximum, solar particles stir up the gases in Earth's upper atmosphere. This raises the level of the atmosphere above the ISS. So, even though there are more solar particles at solar maximum, the ISS is a little better protected by Earth's own atmosphere.

When the Sun is less active, the gases in Earth's atmosphere are lower, and there is not much gas above the ISS to shield it. During these times, scientists lower the orbit of the ISS so it dips into the atmosphere and is more shielded. (Because the atmosphere creates "drag" on the station, they raise the orbit again when possible.)

Sometimes the crew need to go outside the station to make repairs or set up experiments. This is called an Extra Vehicular Activity (EVA). Space suits are designed with as much shielding as possible, but of course the crew are not as well protected during an EVA as when they are inside the station. That's why EVAs are carefully planned to avoid times of high solar activity. Also, the planners try not to put the crew out in their space suits when the station is passing through an area called the South Atlantic Anomaly (ah-NOM-ah-lee). An anomaly is something unusual. The South Atlantic Anomaly is an area above

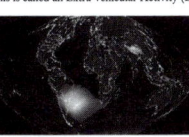
The South Atlantic Anomaly

continued on next page

Name:_____

Space Weather and the International Space Station—Student Reading (continued)

...rth's magnetic field bulges down lower than usual. When the ..., it is not as well shielded by the magnetosphere. Since there are ... area, planners try to keep the crew inside when passing through ...ly.

...rld space agencies constantly watch the Sun's activity. They use ... on spacecraft. If they see a big solar flare, and if they predict ...d toward Earth, they can tell the crew of the ISS to shield ... happen often, but when it ...ncy plans.

...s will be passing through ...go to a small tunnel in the ...nnel has extra shielding

...les is predicted to last a ...lan. The crew builds a ...ers. Water on board the ... (10–12 gallon) containers, ...duffel bag. The containers ... little "house" around the ...rogen in water acts as a

...nd outside the ISS, ...nstantly monitor all UV, ... other energies from the ...odel human "torsos" with ...asure the effects of solar ...mes the torsos are placed ...mes inside. Scientists ...bout how to keep people

Fred, the Human Torso

...the engineering staff of NASA's Johnson Space Center in Houston, Texas, for ...the International Space Station.

SESSION 1.8   Living with a Stormy Sun

Taking the Post-unit 1 Questionnaire

1. **About the questionnaire.** Tell students that they are going to fill out the same questionnaire they filled out at the beginning of the unit. Although the questions are the same, they will be in a different order. Say that they now know much more about the Sun than they did before, and the questionnaire is an opportunity for them to share what they have learned.

2. **Pass out the Post-unit 1 Questionnaire.** Give students about 20 minutes to answer the questions before collecting the questionnaires.

3. **Students as solar-science experts.** Congratulate the students on becoming much more knowledgeable about the Sun and its effects. Say that while most people know about the beneficial effects of the Sun, not many are aware of some of the Sun's harmful effects. Students have become solar-science experts!

One teacher said, "There was a TON of content knowledge in this unit, but it was presented in a way that students understood and, even at sixth grade, I believe they 'got' what they were supposed to get. This area of science is full of misconceptions and this unit makes the information real so the kids understand."

Another said, "I would DEFINITELY USE this unit again. It matches very well with our state's science standards and connects to other units we teach. It's a very concrete way of teaching these concepts so kids are interested in them and also retain the information."

TEACHER CONSIDERATIONS

Name:_____

Post-unit 1 Questionnaire:
How Does the Sun Affect Earth?

1. What things are coming toward Earth from the Sun? List as many things as you can, and be as specific as possible. Next to each thing you list, write how it can harm us, or help us, or both.

What comes from the Sun?	How it can harm us, help us, or both?
_____	_____
_____	_____
_____	_____
_____	_____
_____	_____
_____	_____
_____	_____
_____	_____

Name:_____

Post-unit 1 Questionnaire:
How Does the Sun Affect Earth? (continued)

...always the same?
...No
...cific examples to back up your answer.

_____
_____
_____
_____
_____
_____
_____
_____

Name:_____

Post-unit 1 Questionnaire:
How Does the Sun Affect Earth? (continued)

3. Draw the Earth and Sun system.
- Label the Sun and Earth.
- Show sizes and distances from each other.
- Add words or arrows to show how they move.
- Include anything else to show how the Sun and Earth affect each other.

Name:_____

Post-unit 1 Questionnaire:
How Does the Sun Affect Earth? (continued)

...from harmful **electromagnetic energies** from the Sun?
...rrect answers.

...'s atmosphere

...eld of Earth

...from harmful **particles** from the Sun?

...'s atmosphere

...eld of Earth

...est from the Sun to Earth?

UNIT 1 • 259